T0211494

Women in Engineering and Science

Series Editor
Jill S. Tietjen
Greenwood Village, CO, USA

The Springer Women in Engineering and Science series highlights women's accomplishments in these critical fields. The foundational volume in the series provides a broad overview of women's multi-faceted contributions to engineering over the last century. Each subsequent volume is dedicated to illuminating women's research and achievements in key, targeted areas of contemporary engineering and science endeavors. The goal for the series is to raise awareness of the pivotal work women are undertaking in areas of keen importance to our global community.

More information about this series at http://www.springer.com/series/15424

Mary E. Kinsella

Editor

Women in Aerospace Materials

Advancements and Perspectives
of Emerging Technologies

 Springer

Editor
Mary E. Kinsella
Air Force Research Laboratory (retired)
Centerville, OH, USA

ISSN 2509-6427   ISSN 2509-6435   (electronic)
Women in Engineering and Science
ISBN 978-3-030-40781-0   ISBN 978-3-030-40779-7   (eBook)
https://doi.org/10.1007/978-3-030-40779-7

This Springer imprint is published by the registered company Springer Nature Switzerland AG
The registered company address is: Gewerbestrasse 11, 6330 Cham, Switzerland

Preface

This book provides scholarly chapters detailing the research and development of key aerospace materials that have enabled some of the most exciting air and space technologies in recent years. These scientific and engineering successes are shared by the women who accomplished them, those who were in the labs, on the shop floors, or on the design teams contributing to the realization of these technologies. Their work contributes to the world in the challenging and vital field of aerospace materials, and their stories seethe with pride and passion for the opportunity to make these important contributions.

Materials are the basis of new advancements in technology. They play an especially important role in aerospace by providing properties that enable performance in the extreme environments of air and space. From high-strength, high-temperature materials for engines to stiff and lightweight materials for structures or from radiation-hardened materials for electronics to advanced biomaterials, the world of challenge in aerospace materials is vast. Materials themselves include metals, ceramics, polymers, and composites. Aerospace applications include jets, helicopters, and other air vehicles; rockets, satellites, and other space vehicles; all their associated support and subsystems; and interactions with humans and the environment.

The intention of this book is to benefit a number of audiences, including:

- Young women thinking about careers in materials science or engineering or related fields, working in research or aerospace industry
- Women in these fields interested in learning about and connecting with women doing similar work
- Teachers, professors, advisors, and mentors looking for exemplars (i.e., women engineers and scientists in aerospace materials) to bring to the attention of their students and protégés
- People everywhere interested in learning about

  - Leading-edge aerospace materials research
  - Talented researchers dedicated to advancing materials technology for aerospace missions and for the greater good

– The roles of women in aerospace engineering and science and specifically in materials research

The book begins with a summary of pioneering women in aerospace materials and follows up with chapters describing the work of each individual author and her collaborators. A separate section of biographies is also included, highlighting the backgrounds of the authors, their motivations, and their inspirations.

The materials topics featured in this book represent a small sampling of ongoing research in aerospace materials. Effort has been made to present details of selected materials work and describe the relevant potential aerospace applications. Examples from electronics, optical, biological, metal, and composite materials are all included. Various perspectives are represented, such as processing, testing, characterizing, and manufacturing. The reader will be exposed to the breadth of materials disciplines, the depth of ongoing research therein, and the expanse of potential applications in the realm of aerospace.

All of the contributing authors work in aerospace research and development on projects that include an emphasis on materials. Authors were primarily selected on the basis of this work, their interest in the project, and their enthusiasm for contributing to this volume. Invitations to contribute had to be limited due to the far-reaching scope of the topic; therefore, this book is not representative of all women in the field of aerospace materials, nor was it intended to be.

As the editor of this volume, I am pleased to have had the opportunity to contribute to this series publication, to showcase some of the work that has major impact in the materials and aerospace communities, and to work with several outstanding women who are among the most talented engineers and scientists in research and development.

Centerville, OH, USA   Mary E. Kinsella

Contents

Chapter 1
Aerospace Trailblazers

Jill S. Tietjen

1.1 Introduction

Women's contributions to aerospace materials ensure aircraft and rockets that survive the rigors of air and space travel. A woman invented the molecular density switch monitoring the artificial atmospheres that protect electronic equipment and then alerting humans of the need for remedial action. Another woman invented the propulsion system that keeps communication satellites on orbit after she recognized the efficiencies and other benefits of using only one fuel. The "Queen of Carbon Science" made crucial advances in the understanding of the thermal and electrical properties of carbon nanomaterials. A woman in academia has researched advanced structural alloys used in the aerospace, energy, and automotive industries. Women astronauts have worked on the ceramic tiles that form the heat shield for the space shuttle and served as research chemists investigating organic polymers. The women briefly described in this chapter helped pave the way for the chapter authors that follow in this volume.

1.2 Beatrice Hicks (1919–1979)

Beatrice A. Hicks broke new ground for women as an engineer, inventor, and engineering executive. Because of her interest in mathematics, physics, chemistry, and mechanical drawing in high school, she decided to become an engineer. In fact, her interest had been sparked at age 13 when her engineer father had taken her to see the Empire State Building and the George Washington Bridge, and she learned that it was engineers who built such structures. Although her high school classmates and

J. S. Tietjen, P.E. (✉)
Technically Speaking, Greenwood Village, CO, USA

© Springer Nature Switzerland AG 2020
M. E. Kinsella (ed.), *Women in Aerospace Materials*, Women in Engineering
and Science, https://doi.org/10.1007/978-3-030-40779-7_1

some of her teachers tried to discourage her, pointing out that engineering was not a proper field for women, her parents did not stand in her way.

After her high school graduation in 1935, she entered the Newark (New Jersey) College of Engineering. In 1939, she received a B.S. in chemical engineering and took a position as a research assistant at the College. In 1942, she got a job with the Western Electric Company, becoming the first woman to be employed by the firm as an engineer. She worked first in the test set design department and later in the quartz crystal department. An early award citation stated "the quality of her work became legend." She studied at night while employed and, in 1949, earned an M.S. in physics from Stevens Institute of Technology. Subsequently, she undertook further graduate work at Columbia University. Hicks pioneered the theoretical study, analysis, development, and manufacture of sensing devices, patented a molecular density scanner, and developed an industry model for quality control procedures.

She joined Newark Controls Company, the company her father had founded, as chief engineer in 1945. When her father died in 1946, she became vice president as well. Newark Controls Company specialized in environmental sensing devices. In 1955, she bought control of the company and became president. One of the major products of the company at that time was low-water cutoffs and other devices to protect people from their own forgetfulness, often sold through mail-order companies. During her years with the Company, it manufactured specialized electro-mechanical devices such as liquid level controls, pressure controls, and altitude switches for aircraft and space vehicles where extreme reliability under severe environmental conditions was required.

At Newark Controls Company, Hicks was also involved in the design, development, and manufacture of pressure and gas density controls for aircraft and missiles. In 1959, she was awarded patent 3,046,369 for a molecular density scanner or gas density switch (Fig. 1.1). This type of switch is a key component in systems using artificial atmospheres. The gas density switch (today called a gas density sensor or gas density control or gas density monitor) was designed to monitor gas leakage, particularly for artificial atmospheres around electronic equipment. The artificial atmospheres include dry air, nitrogen, sulfur hexafluoride (SF_6), and fluorochemicals. These gases are suitable as insulators and heat dissipators for a broad range of applications including sealed electronic equipment (non-airborne), sealed airborne electronic equipment, power transformers, switchgear, X-ray units, pressurized power cables, waveguides, and coaxial cables.

At the time of her invention, no mechanism existed to monitor the gas leakage and then signal to the equipment operator that action was warranted. Her gas density switch monitors the number of molecules per unit volume and was capable of indicating leakage at a pressure value that varied with temperature. Density, not pressure or temperature, is the important variable in a sealed-in atmosphere, because it determines arc resistance and heat transfer.

Her switch was a unique bellow-type switch. It could be calibrated to protect throughout the entire temperature range. In addition to its function of signaling critical density, the switch could also be used to activate control circuits of systems, e.g., opening a valve or starting a pump.

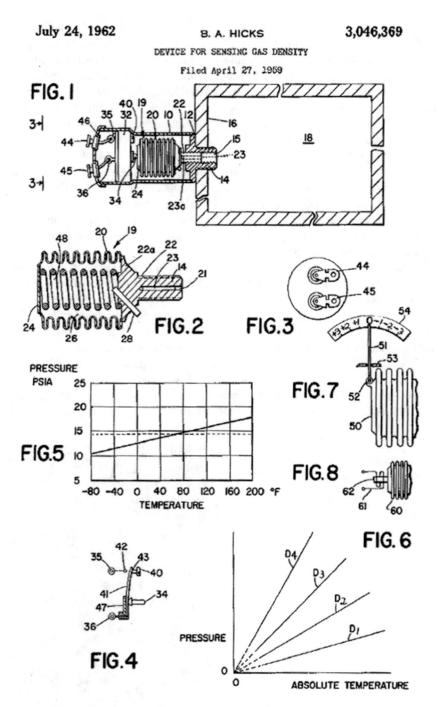

Fig. 1.1 Beatrice Hicks' patent 3,046,369 – gas density sensor. Source: U.S. Patent and Trademark Office

The sub-miniature gas density switch was in use in a number of applications particularly in aircraft and missiles and also in the space program including the Saturn and Apollo missions. The Boeing 707 used a gas density switch to monitor the nitrogen atmosphere around the high voltages in a sealed communications antenna coupler. On both the Boeing 707 and the Hustler B-58, the antenna couplers handle very high voltages and would arc-over in the upper atmospheres. A gas density switch mounted in each antenna coupler indicated remotely at the pilot's panel board if a problem developed.

Fire extinguishers use gas density switches to indicate which units are ready for recharging. Power transformers using gas and vapor cooling are monitored by gas density switches. At the time the switch was developed, it allowed early detection of leakage with an associated remote indicator that prevented critical loss of the gases and permanent damage to the transformers. Gas density switches are of value when power transformers are in shipment or storage, at the time they are put in service, as well as when they are in operation.

Bellow-type gas density switches are still manufactured, in use, and marketed today. One of the most significant applications for these types of gas density switches today is for high-voltage SF_6 breakers. The monitors ensure that the breaker will adequately perform its function of properly interrupting faults. The monitor alarms and then trips the breaker when there is no longer adequate SF_6 to extinguish a fault. Monitors are manufactured in the USA and abroad.

One of the founders of the Society of Women Engineers (SWE), Hicks was elected to serve as its first president in 1950. She was committed to the organization because of her belief that there was a great future for women in engineering. Hicks (Fig. 1.2) received SWE's Achievement Award in 1963 "In recognition of her

Fig. 1.2 Beatrice Hicks – SWE archives, Walter P. Reuther Library, Wayne State University https://reuther.wayne.edu/node/1622

significant contributions to the theoretical study and analysis of sensing devices under extreme environmental conditions, and her substantial achievements in international technical understanding, professional guidance, and engineering education."

Hicks received many other honors and awards over her lifetime. In 1952, she was named "Woman of the Year in Business" by *Mademoiselle* magazine. In 1961, she was the first woman engineer appointed by the US Secretary of Defense to the Defense Advisory Committee on Women in the Sciences. She was the first woman to receive an honorary doctorate from Rensselaer Polytechnic Institute (1965). She also received honorary degrees from Hobart and William Smith Colleges, Stevens Institute of Technology, and Worcester Polytechnic Institute. In 1978, she was elected to the National Academy of Engineering, the sixth woman to be elected. In 2001, she was inducted into the National Women's Hall of Fame. In 2013, Hicks received the Advancement of Invention Award from the New Jersey Inventors Hall of Fame [1–4].

1.3   Yvonne Brill (1924–2013)

Yvonne C. Brill expanded the frontiers of space through innovations in rocket and jet propulsion. Her accomplishments and service had major technical and programmatic impacts on a very broad range of national space programs. Her most important contributions were in advancements in rocket propulsion systems for geosynchronous communication satellites. She invented an innovative satellite propulsion system that solved complex operational problems of acquiring and maintaining station (keeping the satellite in orbit and in position once it is aloft).

Her patented hydrazine/hydrazine resistojet propulsion system (patent no. 3,807,657 – granted April 30, 1974) provided integrated propulsion capability for geostationary satellites and became the standard in the communication satellite industry (Fig. 1.3). Two aspects of Brill's invention are of special significance: she developed the concept for a new rocket engine, the hydrazine resistojet, and she foresaw the inherent value and simplicity of using a simple propellant. Her invention resulted in not only higher engine performance but also increased reliability of the propulsion system and, because of the reduction in propellant weight requirements, either increased payload capability or extended mission life. As a result of her innovative concepts for satellite propulsion systems and her breakthrough solutions, Brill earned an international reputation as a pioneer in space exploration and utilization.

Through her personal and dedicated efforts, the resistojet system was then developed and first applied on an RCA spacecraft in 1983. Subsequently, the system concept became a satellite industry standard. It has been used by RCA, GE, and Lockheed Martin in their communication satellites. The thruster has stood the test of time; more than 200 have been flown. Satellites using her invention form the backbone of the worldwide communication network – 77 of them form the Iridium mobile telephony constellation of satellites, and 54 are geosynchronous

United States Patent [19]

Brill

[11]   **3,807,657**

[45] **Apr. 30, 1974**

[54] **DUAL THRUST LEVEL MONOPROPELLANT SPACECRAFT PROPULSION SYSTEM**

[75] Inventor: **Yvonne Claeys Brill**, Skillman, N.J.

[73] Assignee: RCA Corporation, New York, N.Y.

[22] Filed: **Jan. 31, 1972**

[21] Appl. No.: **221,955**

[52] U.S. Cl. .... **244/1 SB**
[51] Int. Cl. **B64d 3/00**
[58] Field of Search 244/1; 60/200–204, 206–207, 218–220, 224–225, 229, 242

[56]   **References Cited**

UNITED STATES PATENTS

2,968,919	1/1961	Hughes et al.	60/242
3,011,309	12/1961	Carter	60/242
3,015,210	1/1962	Williamson et al.	60/229
3,054,252	9/1962	Beckett et al.	60/203
3,231,223	1/1966	Upper	244/1 SA
3,303,651	2/1967	Grant, Jr. et al.	60/203
3,535,879	10/1970	Kuntz	60/200 R
3,673,801	7/1972	Goldberger	60/218
3,165,382	1/1965	Forte	60/218
3,583,161	6/1971	Simms	60/203

Primary Examiner—Duane A. Reger
Assistant Examiner—Jesus D. Sotelo
Attorney, Agent, or Firm—Edward J. Norton; Joseph D. Lazar

[57]   **ABSTRACT**

A flight auxiliary propulsion system for velocity trim, station keeping, momentum adjustment for a spacecraft comprising rocket or reaction motors, also designated thrusters, utilizing thermally decomposable monopropellants such as hydrazine and other derivatives, thereof hydrogen peroxide, and isopropyl nitrate. The thrusters are arranged in a distribution or manifold system so that one set of thrusters provides for relatively large thrusts of force in the order of 1 to 5 pounds and another set of thrusters develop low thrusts in the millipound range. The large thrusts are developed by the catalytic decomposition of the monopropellant into a thrust chamber and through a throat and expansion nozzle to the ambient externally of the spacecraft. The low level thrusts are developed by heating catalytically or thermally decomposed monopropellant by electrical heating elements more commonly known as resisto-jet elements. Dual thrust levels may also be achieved by a common motor with a controllable resisto-jet and variable throat-area control.

10 Claims, 5 Drawing Figures

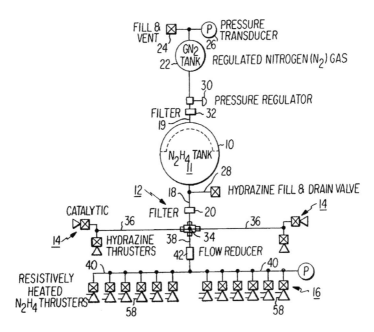

Fig. 1.3 Yvonne Brill patent 3,807,657. (Source: U.S. Patent and Trademark Office)

communication satellites. The impact of global satellite communications extends to all walks of life, from national security to commercial telephone, from remote medicine and education to international trade. Brill's innovation, by enabling a dramatic improvement in satellite capability, has directly improved all of these endeavors. Every time one of our soldiers uses his/her Iridium satellite telephone, that solider is directly benefiting from Brill's innovation.

The invention of the hydrazine/hydrazine resistojet and its extensive use on current communications is just one of the many contributions Brill made to expanding space horizons. Her other significant technical achievements include work on propellant management feed systems, electric propulsion, and an innovative propulsion system for the Atmosphere Explorer, which, in 1973, allowed scientists to gather extensive data of the earth's thermosphere for the first time. Brill defined, successfully advocated, and conducted a program to evaluate capillary propellant management for three-axis stabilized spacecraft. Capillary propellant management is now routinely used on a significant fraction of US space systems. Her system has led to major improvements in the capabilities and competitiveness of very large numbers of US spacecraft.

Brill managed the fabrication, assembly, integration, and test of a complex Teflon solid propellant pulsed plasma propulsion system (TSPPS), also called pulsed plasma thrusters (PPTs). She resolved many technical and design problems in the process of bringing TSPPS from experimental to operational use in satellites, including the NOVA I spacecraft launched in May 1981, which formed part of the US Navy's Navigational Satellite System. Her efforts both provided the solution for an unprecedented navigational capability and opened the way for the now routine use of electric propulsion on commercial Western space systems. In addition, PPTs, which are direct descendants of her design, are now being developed for propulsion functions on small/micro-government spacecraft for many applications.

Brill brought to society the benefits of her bountiful knowledge and wisdom by consulting with governments and space agencies throughout the world. She was instrumental in the success of several satellite system developments for the International Maritime Satellite (INMARSAT) organization and for Telenor, the Norwegian telecommunications organization. She served as one of nine members of NASA's Aerospace Safety Advisory Panel (ASAP) which was created in 1968 after the Apollo 204 Command Module spacecraft fire in 1967 to focus on safety issues. During her period of service, ASAP defined and recommended many technical and programmatic changes to enhance Orbiter safety that were subsequently implemented by NASA. Specific examples of such impacts include a modified Space Shuttle Main Engine (SSME) heat exchanger design that eliminated a catastrophic failure mode, a SSME design change that enabled lowered turbine inlet temperatures with increased safety margins, and increased staffing at NASA space flight centers in support of Orbiter flows.

During her long and stellar career, Brill was a pioneer in the field of space technology. Throughout most of that time, she was the sole technical woman working in rocket propulsion systems. As she excelled in her career, Brill worked tirelessly to nominate women for awards and to boards and served as a role model for several generations of women engineers, including her daughter.

Brill became a member of the National Academy of Engineering in 1987 and was a fellow of SWE and the American Institute of Aeronautics and Astronautics. Among her many awards were the 1986 SWE Achievement Award "for important contributions in advanced auxiliary propulsion of spacecraft and devoted service to the growing professionalism of women in engineering," the 1993 SWE Resnik Challenger Medal for expanding space horizons through innovations in rocket propulsion systems, and induction into the Women in Technology International Hall of Fame in 1999. After her induction into the New Jersey Inventors Hall of Fame (the first woman) in 2009 and the National Inventors Hall of Fame in 2010, Brill received the nation's highest honor, the National Medal of Technology and Innovation, from President Obama in 2011 (Fig. 1.4) "For innovation in rocket propulsion systems for geosynchronous and low earth orbit communication satellites, which greatly improved the effectiveness of space propulsion systems" [4–7].

Fig. 1.4 Yvonne Brill with her National Medal of Technology and Innovation poster and medal. (Source: Author)

1.4   Mildred Dresselhaus (1930–2017)

At the Massachusetts Institute of Technology for nearly 60 years, Dr. Mildred Dresselhaus was the first female recipient of the National Medal of Science in the engineering category, the first woman to receive the Institute of Electrical and Electronics Engineers (IEEE) Medal of Honor, and was known as the "Queen of Carbon Science." She was the first woman tenured in the School of Engineering at MIT. In August 2000, she became the Director of the Office of Science in the Department of Energy, having been nominated by President Clinton and confirmed by the US Senate.

Dresselhaus was a solid-state physicist and materials scientist whose research areas included superconductivity; the electronic and optical properties of semimetals, semiconductors, and metals; and, particularly, carbon-based materials. She made crucial advances in the understanding of the thermal and electrical properties of carbon nanomaterials. She was the first female institute professor at MIT, an honor that recognizes distinguished accomplishments in scholarship, education, service, and leadership. Her professorship was in electrical engineering and physics.

Her other honors demonstrate the recognition she received for her significant accomplishments. The citation for her 1990 National Medal of Science reads "For her studies of the electronic properties of metals and semimetals, and for her service to the Nation in establishing a prominent place for women in physics and engineering." Her 2014 US Presidential Medal of Freedom (the highest honor for civilians) citation reads for "deepening our understanding of condensed matter systems and the atomic properties of carbon – which has contributed to major advances in electronics and materials research." On presenting her the award, President Obama said "Her influence is all around us, in the cars we drive, the energy we generate, the electronic devices that power our lives."

In 2017, she received the Benjamin Franklin Medal "For her fundamental contributions to the understanding and exploitation of carbon nanomaterials, such as the spheres known as Buckminsterfullerenes, the cylindrical pipes called nanotubes, and the single-atom-thick sheets of carbon known as graphene, and for launching the field of low-dimensional thermoelectricity, the direct conversion of heat to electricity." She was the first solo recipient of the Kavli Prize for her pioneering contributions to the study of phonons, electron-phonon interactions, and thermal transport in nanostructures.

Growing up poor but with exceptional musical ability, Dresselhaus received a scholarship to music school. Determined to pursue every education opportunity she could, she applied to Hunter College High School for Girls. There, she received encouragement to study physics at Hunter from her advisor Rosalyn Yalow (later a Nobel Laureate) as opposed to becoming a schoolteacher. Dresselhaus received an A.B. from Hunter College in 1951 in physics and math. After a year in Cambridge, England, on a Fulbright scholarship in physics, she studied first at Harvard, completing her master's degree and then received her Ph.D. from the University of

Chicago in 1958 with a thesis on superconductors. She then served as National Science Foundation postdoctoral fellow at Cornell University.

In 1960, Dresselhaus joined the Lincoln Lab at MIT, where she studied the properties of graphite. Although her early results were not encouraging, she persevered to obtain data that provided the most accurate characterization of carbon's electronic band structure that had ever been achieved. As a result, she was appointed as a visiting faculty member in the Electrical Engineering Department at MIT under the Abby Mauze Rockefeller Fund, established to promote the scholarship of women in science and engineering. Soon, she achieved tenured professorship. When this occurred in 1968, she became MIT's first tenured female faculty member in engineering.

During the 1970s, Dresselhaus studied graphene intercalation compounds and built her reputation. Her work resulted in a better understanding of fundamental quantum concepts which Dresselhaus used to update theoretical equations as they apply to nonmaterial systems.

A decade later, Dresselhaus's group pursued new carbon materials by blasting graphite with lasers. The ablation produced large carbon clusters of 60 or 70 atoms. Richard Smalley, who was independently performing similar experiments, identified the clusters as fullerenes, more commonly called buckyballs. Dresselhaus spent the remainder of her career calculating the intricacies of carbon nanotubes, which she determined could be formed by elongating the fullerene structure instead of terminating it into a buckyball. She showed that the electrical properties of the carbon nanotubes changed with the orientation of the hexagonal structure. Her calculations of these electrical properties revealed that carbon nanotubes could be applied as either metals or semiconductors.

The 1977 recipient of SWE's Achievement Award "for significant contributions in teaching and research in solid state electronics and materials engineering," Dresselhaus (Fig. 1.5) co-founded the MIT Women's Forum in 1970. The Forum

Fig. 1.5 Mildred S. Dresselhaus – SWE archives, Walter P. Reuther Library, Wayne State University http://reuther.wayne.edu/node/1527

met weekly and provided a venue where women could discuss issues relevant to work and life. The objective of the forum was to support the careers of women in science and engineering at MIT. In 1999, she received the Nicholson Medal for Humanitarian Science from the American Physical Society "for being a compassionate mentor and lifelong friend to young scientists; for setting high standards as researchers, teachers and citizens; and for promoting international ties in science."

In addition to her many honorary degrees, Dresselhaus served as president of the AAAS, president of the American Physical Society, a member of the National Academy of Engineering and the National Academy of Sciences, and a fellow of SWE, AAAS, IEEE, and others. Dresselhaus co-authored 8 books and about 1700 papers and supervised more than 60 doctoral students. She demonstrated a strong commitment to promoting gender equity in science and engineering. In describing the importance of persistence, Dresselhaus said "It was what you did that counted and that has followed me through it" [5, 8–21].

1.5   Bonnie Dunbar (1949–)

Astronaut Dr. Bonnie J. Dunbar is another pioneering engineering woman. When she enrolled as an engineering student at the University of Washington, there were nine women in her entire freshman class. She received B.S. and M.S. degrees in ceramic engineering from the University of Washington in 1971 and 1975, respectively. When she joined the Astronaut Corps in 1980, she was in only the second class at NASA to accept women. Subsequently, she earned her Ph.D. in mechanical/biomedical engineering at the University of Houston in 1983 where her dissertation focused on the evaluation of the effects of simulated bone strength and fracture toughness.

Prior to becoming an astronaut, Dunbar was employed at Boeing Computer Services as a systems analyst for 2 years before conducting the research for her master's degree. She then worked as a senior research engineer at Rockwell International Space Division, where she played a key role in the development of the ceramic tiles that form the heat shield for the space shuttle, allowing it to reenter the earth's atmosphere. She helped establish the equipment, process, and standards to manufacture the tiles. In 1978, Dunbar became a payload officer and flight controller for NASA. She served as a guidance and navigation officer and flight controller for the Skylab reentry mission in 1979. Later, she worked with the thermal tile repair team to analyze the tile problem on *Columbia* and explore possibilities for in-orbit tile repair.

While an astronaut, Dunbar's NASA technical assignments included verification of shuttle flight software; serving as a member of the Flight Crew Equipment Control Board; 13 months in training in Star City, Russia, for a 3-month flight on the Russian space station, Mir; and assistant director with a focus on university research. She has logged more than 50 days in space and flew on 5 flights. Dunbar's experiments in space have involved protein crystal growth; surface tension physics;

and tests on muscle performance, bones, the immune system, and the cardiopulmonary system.

Dunbar served as assistant director at the Johnson Space Center and later deputy associate director for Biological Sciences and Applications before her retirement from NASA. In 2005, she became president and chief executive officer of the Seattle Museum of Flight in Washington State. In 2016, she joined Texas A&M where she serves as The Engineering Experiment Station (TEES) eminent professor and director of the Aerospace Human Systems Laboratory. Her current research interests include microgravity and partial g fluid physics, digital human modeling, extravehicular activity spacesuit system design, space systems engineering, and engineering education.

Dunbar (Fig. 1.6) received the Resnik Challenger Medal from SWE in 1992 which acknowledges a specific engineering breakthrough or achievement that has expanded the horizons of space exploration. She received the IEEE Judith Resnik Award in 1993 "for outstanding contributions to space engineering." In 2005, she received SWE's Achievement Award "In recognition of her visionary contributions ranging from ceramic shuttle-tile design to biomedical research. Dr. Dunbar's efforts benefit astronautics, humankind, and the future scientists and engineers she

Fig. 1.6 Bonnie Dunbar –
SWE archives, Walter
P. Reuther Library, Wayne
State University http://
reuther.wayne.edu/
node/1532

inspires." She has been inducted into the Women in Technology International Hall of Fame. Dunbar was elected to the National Academy of Engineering and the ASF Astronaut Hall of Fame and is a fellow of the American Institute of Aeronautics and Astronautics and the American Ceramic Society. She received the American Association of Engineering Societies' Norm Augustine Award for Outstanding Achievement in Engineering Communications and the American Society of Mechanical Engineers' Ralph Roe Award. Dunbar has been elected a fellow of the Royal Society of Edinburgh and is a fellow of the Royal Aeronautical Society [22–25].

1.6   Tresa Pollock (1961–)

Currently the Department Chair and Alcoa Distinguished Professor of Materials at the University of California, Santa Barbara, Tresa Pollock started her undergraduate education at Purdue University as a first-generation college student. She chose to study what was called metallurgical engineering at the time because she wanted to help solve technological and societal problems. Today, she is known worldwide for her expertise in advanced structural alloys with applications in the aerospace, energy, and automotive industries and as a teacher and mentor.

Her life's work was set into motion during her college years when Pollock co-oped at Allison Gas Turbines (later Rolls-Royce). Her research almost 40 years later still focuses on aircraft engine alloys and the needs of the aerospace industry as well as structural materials challenges in the automotive and energy generation sectors. Pollock says, "Collaboration among fields for the solution of complex, multidisciplinary problems is exciting for me. We have some big challenges in the next couple of decades in energy, transportation, the environment, climate, space, genetics, and healthcare, to name a few."

In addition to her B.S. from Purdue University, Pollock earned her Ph.D. in Materials Science and Engineering from MIT. She began her professional career at GE Aircraft Engines, where she worked in the development of advanced superalloys for gas turbine engines. The allure of academia was strong, and, in 1991, she joined the faculty at Carnegie Mellon University in the Department of Materials Science and Engineering as the Alcoa professor. In 2000, she joined the Materials Science and Engineering Department at the University of Michigan where she served until moving to the Materials Department at the University of California, Santa Barbara. Over her career, she has worked in the fields of metallurgy, microstructural characterization, materials science, and nickel-based superalloys.

Her current research interests include:

- Mechanical and environmental performance of materials in extreme environments
- Unique high-temperature material processing paths
- Ultrafast laser-material interactions
- Alloy design

- 3-D material characterization
- Thermal barrier coating systems and platinum group metal-containing bond coats
- New intermetallic-containing cobalt-based materials
- Vapor phase processing of sheet materials for hypersonic flight systems
- Growth of nickel-based alloy single crystals
- Development of new femtosecond laser-aided 3-D tomography techniques
- Development of models for Integrated Computational Materials Engineering efforts

The work in her laboratory on high-temperature materials for jet engines – particularly nickel-based superalloys – has already helped improve the safety and efficiency of almost every jet turbine engine. Aircraft engine materials can reach such extreme temperatures during takeoff that they almost melt! Alloy materials thus must be developed that operate at 90% of their melting temperature while simultaneously bearing significant mechanical loads. Pollock and her team make materials that matter – making aircraft possible, allowing efficient high-temperature operation for space vehicles as well as power plants. The materials for energy generation affect thermoelectrics, fuel cells, and energy plants fueled with natural gas as well as alternative fuels and include light alloys that have a high degree of resistance. She also works on protective coatings that have been engineered to withstand extreme environments. Her work in the area of integrated material computation targets bond strength and chemical potential examining the absence of cracks and the crack resistance of specific materials. She aims to identify optimized components, a more efficient development process, and economic and efficient manufacturing processes.

Growing up in Ohio near Wright-Patterson Air Force Base, Pollock was enchanted early on with aerospace, astronauts, and Amelia Earhart. She says "You can't be afraid to do something difficult. You have to be adventurous and not be doing the same that everybody else is." She also says, "You have to be comfortable with the fact that you do not know everything. You never will. In doing that, it helps you reach out across boundaries to work with other people."

Her many honors and awards include election to the National Academy of Engineering, a fellow of ASM, and fellow of The Minerals, Metals and Materials Society (TMS) "for seminal contributions in the understanding of high temperature alloys, and for distinguished leadership in materials education and the materials profession." Her awards recognize her contributions to the literature, excellence in teaching, as well as overall professional accomplishment. She is a member of the German National Academy of Sciences Leopoldina.

Pollock believes her biggest contribution to the world is not from the research she has conducted but from the students she has in her laboratory and what they will accomplish [26–30].

1.7   Cady Coleman (1960–)

When she was 19, Cady Coleman wanted to study chemistry. She decided to go to MIT because of an experience she had in a physics class in high school. She relates, "I was in a physics class with just a few guys and they all wanted to go to MIT – and it was really clear that it never occurred to them that I might want to go, too." So, she applied just to prove that she could get in – and she was accepted. Although her previous impression was that MIT was a "nerdy, one-dimensional school," during her visit to campus, she found well-rounded students who were like her, and she decided to attend.

Although she had previously had no plans to be an astronaut, while Coleman was an undergraduate student at MIT, astronaut Sally Ride, Ph.D., came and gave a talk. The opportunity to hear Ride, talk to her, and see her passion changed the direction of Coleman's life. Coleman says, "It had never occurred to me to do that kind of job until I met her and saw somebody who had a job where it really counted that they were well-educated and passionate about what they were studying and at the same time there was this sort of adventure in their lives."

Coleman did complete her B.S. degree in chemistry at MIT in 1983 on her ROTC scholarship and joined the US Air Force. She received her doctorate in polymer science and engineering from the University of Massachusetts in 1991. She applied to and was selected by NASA for the Astronaut Corps in 1992.

During her years with the Air Force, Coleman worked as a research chemist at Wright-Patterson Air Force Base where she investigated the use of organic polymers in applications including advanced computers and data storage. She has served as a surface analysis consultant for the Long Duration Exposure Facility launched from one space shuttle mission and retrieved 6 years later by another.

Coleman was a "human guinea pig" for the centrifuge program at the Crew Systems Directorate of the Armstrong Aeromedical Laboratory setting endurance and tolerance records as she participated in physiological and new equipment studies. She also served as the Chief of Robotics for the Astronaut Office which included robotic arm operations and training for all space shuttle and International Space Station missions. In 2004, she lived underwater for 11 days as an aquanaut during the NEEMO 7 (NASA Extreme Environment Mission Operations) mission in the Aquarius Underwater Laboratory.

Coleman flew on space shuttle and Soyuz missions and spent more than 5 months on the International Space Station. Her first flight was in 1995 aboard *Columbia*. That mission focused on materials science, biotechnology, combustion science, the physics of fluids, and science experiments in the Spacelab module. In 1999, she served as mission specialist on the Space Shuttle *Columbia* and was in charge of launching the Chandra X-ray Observatory and its Inertial Upper Stage from the shuttle's cargo bay. Aboard the ISS, Coleman was the lead science and lead robotics officer performing osteoporosis experiments examining liquid behavior and conducting robotics experiments.

Coleman now spends a significant part of her time encouraging people to pursue careers in science, technology, engineering, and mathematics. She says, "I feel strongly about encouraging women and minorities to be in STEM. I think so much of what made it possible for me to be selected for this job and to have the courage to stand up and ask for it came from those early days, and I wanted to say 'thank you' . . . I really think there's still a great need for young women and minorities to see people they can identify with just to cement that theory that 'this could be you.' There's always the theory that everybody can do anything, but it's about making that theory feel really real, and I think that meeting someone or seeing them in person can have that effect. It did for me" [31–34].

References

1. P. Proffitt (ed.), *Notable Women Scientists* (Gale Group, Farmington Hills, 1999)
2. M. D. Candee (ed.), *Current Biography* (The H. W. Wilson Company, New York, 1957)
3. Beatrice Hicks recognized by New Jersey Inventors Hall of Fame, SWE Magazine, Winter 2014, http://www.nxtbook.com/nxtbooks/swe/winter14/index.php#/22. Accessed 24 May 2015
4. A. Stanley, *Mothers and Daughters of Invention: Notes for a Revised History of Technology* (Rutgers University Press, New Brunswick, 1995)
5. www.swe.org/SWE/Awards,achieve3.htm. Accessed 1 Sept 1999
6. Yvonne Brill, www.witi.org/center/witimuseum/halloffame/1999/ybrill.shtml. Accessed 14 Feb 2001
7. President Obama honors nation's top scientists and innovators, https://www.whitehouse.gov/the-press-office/2011/09/27/president-obama-honors-nation-s-top-scientists-and-innovators. Accessed 25 May 2015
8. S. Ambrose, K. Dunkle, B. Lazarus, I. Nair, D. Harkus, *Journeys of Women in Science and Engineering: No Universal Constants* (Temple University Press, Philadelphia, 1997)
9. *Who's Who in Technology*, 7th edn. (Gale Research, Inc., New York, 1995)
10. Briefings: Medal of Honor Goes to Dresselhaus, *The Institute*, March 2015
11. M. Anderson, The queen of carbon, *The Institute*, May 2015
12. Award Recipient, www.interact.nsf.gov/MOS/Histrec.nsf/. . . Accessed 14 Feb 2001
13. 1999 Nicholson Medal for Humanitarian Service to Mildred S. Dresselhaus MIT, www.aps.org/praw/nicholso/99wind.html. Accessed 14 Feb 2001
14. Mildred S. Dresselhaus – 1997 AAAS President, www.aaas.org/communications/inside17.htm. Accessed 14 Feb 2001
15. Mildred Spiewak Dresselhaus, www.witi.org/center/witimuseum/halloffame/1998/mdresselhau.shtml. Accessed 14 Feb 2001
16. S.J. Ortiz, View from the inside: Meet Mildred Dresselhaus: New Director of the Office of Science, www.pnl.gov/energyscience/08-00/inside.htm. Accessed 14 Feb 2001
17. MIT News Office, Institute Professor Emerita Mildred Dresselhaus, a pioneer in the electronic properties of materials, dies at 86, http://news.mit.edu/2017/institute-professor-emerita-mildred-dresselhaus-dies-86-0221. Accessed 7 May 2019
18. The Franklin Institute Awards, Mildred S. Dresselhaus, https://www.fi.edu/laureates/mildred-s-dresselhaus. Accessed 7 May 2019
19. The Kavli Prize, Mildred Dresselhaus, http://kavliprize.org/prizes-and-laureates/laureates/mildred-s-dresselhaus. Accessed 7 May 2019
20. APS News, Mildred Dresselhaus: 1930–2017, https://www.aps.org/publications/apsnews/updates/mildred.cfm. Accessed 7 May 2019

21. The Heinz Awards, 11th, Mildred Dresselhaus, http://www.heinzawards.net/recipients/mildred-dresselhaus. Accessed 7 May 2019
22. Dr. Bonnie J. Dunbar, www.witi.com/center/witimuseum/halloffame/2000/bdunbar.shtml. Accessed 22 Dec 2000, pp. 1–2
23. Biographical Data, www.jsc.nasa.gov/Bios/htmlbios/dunbar.html, accessed December 22, 2000, pp. 1–2
24. B.J. Dunbar, College of Engineering, Texas A&M University, https://engineering.tamu.edu/aerospace/profiles/dunbar-bonnie.html. Accessed 16 May 2019
25. B.J. Dunbar., http://www.astronautix.com/d/dunbar.html. Accessed 16 May 2019
26. MRL: Materials Research Laboratory at UCSB: An NSF MRSEC, Tresa Pollock, https://www.mrl.ucsb.edu/people/faculty/tresa-pollock. Accessed 21 Apr 2019
27. T. Pollock., https://en.wikipedia.org/wiki/Tresa_Pollock. Accessed 22 Apr 2019
28. P. Plackis-Cheng, Making materials that matter, http://www.impactmania.com/article/tresa-pollock-making-materials-that-matter/. Accessed 22 Apr 2019
29. T.M. Pollock, Purdue University College of Engineering, 2008 DEA, https://engineering.purdue.edu/Engr/People/Awards/Institutional/DEA/DEA_2008/pollock. Accessed 22 Apr 2019
30. Curriculum Vitae Prof. Dr. Trersa Pollock, Leopoldina National Akademis der Wissenschaften, https://www.leopoldina.org/fileadmin/redaktion/Mitglieder/CV_Pollock_Tresa_EN.pdf. Accessed 16 May 2019
31. R. Kirkland, Former NASA Astronaut Cady Coleman to Speak at STEM Symposium, Washington Exec, April 2018, https://washingtonexec.com/2018/04/former-nasa-astronaut-cady-coleman-to-speak-at-stem-symposium/. Accessed 22 Apr 2019
32. C. Coleman., https://en.wikipedia.org/wiki/Catherine_Coleman, Accessed 22 Apr 2019
33. NASA Astronaut Cady Coleman Joins Skycatch Board of Directors, https://blog.skycatch.com/nasa-astronaut-cady-coleman-joins-skycatch-board-of-directors, August, 13, 2018. Accessed 22 Apr 2019
34. Coleman, Catherine Grace 'Cady', www.astronautix.com/c/colemancatherine.html. Accessed 22 Apr 2019

Chapter 2
Peeking Inside the Black Box: NMR Metabolomics for Optimizing Cell-Free Protein Synthesis

Angela M. Campo, Rebecca Raig, and Jasmine M. Hershewe

2.1 Literature Review of Cell-Free Protein Synthesis

CFPS utilizes the transcription and translation hardware of cells to produce specific proteins encoded by a template DNA that is provided to reactions. Expression of desired proteins can be accomplished in living cells, but the production of proteins and metabolites necessary to sustain life can compete with the production of the desired protein. CFPS offers an efficient means to produce a protein or chemical of interest, even if that product is toxic to a living cell. Cell-free systems have been used in biochemistry for many years; in fact, the first cell-free experiment was reported by Eduard Buchner in 1897 [1]. Buchner's work is considered by many to be the dawn of biochemistry [2].

Eduard Buchner was a German chemist who won the 1907 Nobel Prize in Chemistry for discovering that fermentation is an enzyme-driven process [3]. Before Buchner's work, there was debate in the scientific community surrounding the mechanism of fermentation. The dominant group of scientists believed in a vital force that exists in all living organisms, and that vitalism was necessary for biological

A. M. Campo (✉)
Materials and Manufacturing Directorate, Air Force Research Laboratory,
Wright-Patterson Air Force Base, OH, USA
e-mail: angela.campo@us.af.mil

R. Raig
Air Force Research Laboratory, Materials and Manufacturing Directorate,
Wright-Patterson Air Force Base, Dayton, OH, USA

UES, Inc., Dayton, OH, USA
e-mail: Rebecca.raig.ctr@us.af.mil

J. M. Hershewe
Jewett Lab, Department of Chemical and Biological Engineering, Northwestern University,
Evanston, IL, USA
e-mail: jasminehershewe2016@u.northwestern.edu

© Springer Nature Switzerland AG 2020   19
M. E. Kinsella (ed.), *Women in Aerospace Materials*, Women in Engineering
and Science, https://doi.org/10.1007/978-3-030-40779-7_2

processes [4] instead of a substrate-driven process dependent on enzymes. The opposing scientists resisted the belief in vitalism but had struggled to provide proof to solidify their theory.

Buchner's Nobel work resulted from unexpected fermentation of "press juice," the paste that results from grinding yeast. While preparing press juice for an experiment, he mixed it with a sugar solution intending to preserve the juice, but instead he soon noticed the signs of fermentation [5]. Buchner harnessed the enzymes in the yeast to perform fermentation in a cell-free manner. This serendipitous observation was the basis for his future Nobel work. He was able to prove that cane sugar could be fermented with press juices. This discovery shocked the vitalism supporters; until Buchner's experiment, they had believed that only living organisms could conduct fermentation. Eduard Buchner's experiments revealed the complex world of metabolism and enzyme-catalyzed reactions, which ushered in the era of modern biochemistry.

Once Buchner revealed cell-free systems as a research tool, other scientists exploited these systems to study complex biological mysteries. Cell-free systems were famously used to study protein transcription and translation. DNA was suspected to be the genetic material in a cell, but it was unclear how DNA could lead to protein production. Marshall Nirenberg and J. Heinrich Matthaei created synthetic RNA fragments that consisted of repeated units of one nucleotide. When they inserted these synthetic RNA fragments into a crude cell-free system, a single protein was produced. For example, when they inserted RNA that consisted of repeated uracil-containing nucleotides, this resulted in the production of a protein that contained only the amino acid phenylalanine [6]. This elegant but simple experiment was the Rosetta Stone which enabled the genetic code to be broken.

The Nirenberg and Buchner experiments are just a few examples of cell-free protein synthesis reactions that have contributed to science. As the technique was refined, CFPS research attained significant milestones in more efficient preparation, increased yields, and pertinence to numerous applications. Figure 2.1 shows a timeline of selected CFPS advances [7–11].

CFPS has the potential to produce a wide assortment of materials from proteins to small molecules of interest [12]. The aerospace materials community has a significant challenge that will need to be addressed in the near future; how do we continue to produce fossil-fuel based materials? Fossil fuels are the backbone of many aerospace materials such as hydraulic fluid [13], coolants, and fuels [14]. The world reserves of fossil fuels are expected to be depleted within the next 50–120 years

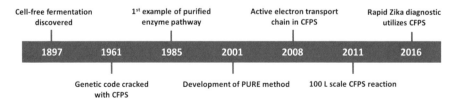

Fig. 2.1 Highlights in CFPS advancements

[15], which will force materials researchers to adapt and develop new techniques to produce oil-dependent products. Cell-free protein synthesis may be able to fill this technology gap. CFPS can be employed to synthesize a wide variety of commodity chemicals, far more than just proteins. Bacteria endogenously produce a variety of alcohols and acids which could be utilized as fuels. Natural pathways to produce chemicals of interest can be bioengineered into a bacterial chassis to allow production of a chemical that is not produced endogenously by the host organism or is produced at low levels [16, 17]. Using a combined CFPS and metabolic engineering approach, enzymes that perform biosynthetic transformations to make a small-molecule product can be rapidly expressed and optimized [18]. Additionally, since CFPS is amenable to liquid handling, it is a powerful tool for testing synthetic or uncharacterized enzyme pathways for bioproduction of useful molecules and metabolites [19, 20]. It is notable here that several non-model cell-free systems (ranging from mammalian systems to yeast and non-model bacteria) have been developed for prototyping enzymes in context other than *E. coli* lysates, expanding the flexibility of CFPS for producing products [19, 21–23].

1-Butanol is a molecule that has the potential to replace diesel gasoline fuels in the future. 1-Butanol has an energy density of 29.2 MJ/L, which is comparable to gasoline at 32.0 MJ/L [24]. 1-Butanol has a lower energy density than diesel fuel, which has an energy density of 35.86 MJ/L [24], but has many other desirable fuel characteristics such as higher viscosity and lubricity than other alternative fuels [24], is miscible with diesel and gasoline [25], and would be compatible with existing fuel distribution infrastructure [26] in part due to its hydrophobic nature. Combustion experiments performed in a heavy-duty diesel engine found that soot and CO emissions were significantly reduced when a diesel-butanol fuel blend was used [27]. Butanol can be produced from multiple feedstocks [28] via acetone-butanol-ethanol (ABE) fermentation with a living microbial chassis, but the yield is hindered due to the eventual toxicity of the butanol to the chassis. This presents an opportunity for CFPS, since the toxicity is not an issue because the lysate is not a living organism.

1-Butanol could be a fuel source for aircraft maintenance equipment such as hydraulic fluid service carts and generators. Supplying energy to forward operating bases (FOBs) is an expensive and dangerous endeavor due to the rough and contested terrain that must be traversed. In 2007, a report estimated that one casualty occurred for every 24 fuel convoys [29] and between 70% and 90% of the convoys [30] to FOBs are fuel supply missions. The estimated costs for fuel delivery can range from $10 to 50 per gallon; if an airdrop is necessary, the cost soars to over $400 per gallon [30]. The remote production of butanol at FOBs could drastically reduce supply missions, saving lives and reducing cost. On-site butanol production has an additional benefit in that it could be designed to utilize waste materials as carbon substrates. Godoy et al. [31] have found that the triglycerides present in urban wastewater treatment sludge can be converted to ethanol via a genetically modified microbe. This genetic circuit could be modified to produce butanol in favor of ethanol.

CFPS could be the key not only to energy production for military bases but also to recycling the massive amounts of trash that are generated on base. The Department of Defense (DoD) conducted a study in 2011 and determined that in-theater bases accumulate 8–10 pounds of waste per day per person, totaling 60,000–85,000 pounds of solid waste generated per day at a large base [32]. The waste situation is unsustainable and has led to the use of burn pits in the past [33]. The use of incinerators at FOBs does not fully solve the problem as some FOBS are "black-out" bases and cannot use incinerators that would generate light during the night, as this would provide a target to insurgents for rocket attacks [34]. Fiberboard accounts for 40.9% of the military waste stream with food waste at 11.2% [35]. CFPS could be optimized to utilize the food and fiberboard waste as carbon sources to produce butanol or other commodity chemicals of interest. CFPS has the potential to cut military waste in half while producing a needed material.

Before CFPS can be utilized to produce aerospace materials, the inherent variability in CFPS reactions must be understood, and a path to mitigate this variability needs to be developed. CFPS reactions consist of a series of complex biochemical processes that must occur to produce the desired product. In traditional organic synthesis, reactions are typically conducted in discrete steps. CFPS reactions have multiple events simultaneously occurring, which complicates metabolic monitoring. Figure 2.2 is a general schematic of cell-free protein synthesis reaction components and describes the metabolic and biosynthetic events that must occur for CFPS to successfully produce a product. Briefly, reaction components consist of an S30 extract from *E. coli* cells, which contain ribosomes and the soluble components of the cells. Prior to preparing the extract, cells are lysed, and the cell wall and genomic DNA are removed from the extract via centrifugation. Energy to fuel the biosynthetic

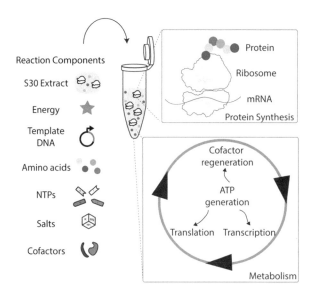

Fig. 2.2 Cell-free protein synthesis reaction components and process overview

reactions, a template DNA encoding for the protein of interest, amino acids, nucleotide triphosphates, salts, and necessary cofactors are also supplemented to reactions. More information about these components can be found in the Experimental section. Transcription of the template DNA results in mRNA templates that can be translated to protein product by ribosomes in the extract. Importantly, metabolism in the extract is active for cofactor regeneration and ATP generation.

Given this complex situation, with many competing events and side reactions occurring at the same time, CFPS can be difficult to optimize. CFPS reactions have been compared to a black box, where reactants are inserted, and the product is formed at varying yields with the intervening steps shrouded in mystery. Although the basic pathways are known, these complex reactions can follow unexpected pathways that divert protein production. By employing in vitro NMR spectroscopy, we can monitor CFPS reactions to understand the evolution of substrates and products over time. Elucidating these processes will enhance our understanding of, and ability to optimize, CFPS and cell-free biology.

2.2 Experimental

Monitoring CFPS reactions can be accomplished by multiple methods. Commonly, CFPS reactions are supplied with a template DNA encoding for a fluorescent protein, and then reactions are monitored with a plate reader. CFPS reactions can be monitored by mass spectroscopy (MS), but this would require either replicates flash frozen at each time point or aliquots taken from the reaction. Taking aliquots is undesirable as this would change the volume for the reaction, which directly impacts the yield as it changes the surface area to volume ratio. CFPS reactions rely on oxygen to conduct oxidative phosphorylation that produces ATP in the lysate. MS analysis is also complicated by the complex nature of the cell lysate, difficulty in analyzing the m/z data, and the time required to process each sample and analyze results. Therefore, NMR has several important advantages for monitoring and studying the core biological processes of CFPS reactions. The CFPS reaction can be conducted in an NMR tube, providing a mechanism to monitor the reaction, while it proceeds without needing to take aliquots or prepare multiple replicates for MS analysis. NMR spectroscopy is non-destructive, is highly repeatable in identifying chemical compounds, can be tailored to exclude undesirable signals such as water and/or high molecular weight compounds, has simple sample preparation, and enables in vitro metabolomics profiling. In this, we monitor CFPS productivity using a superfolder green fluorescent protein (sfGFP) reporter. In tandem, reactions from the same CFPS mix are monitored via NMR to obtain metabolite profiles over time (Fig. 2.3). Fluorescence is typically tracked using a plate reader at the 5–50 μL volume, while NMR studies are conducted at the 200 μL scale. Reactions are monitored for 3 hours, which is a timeframe where CFPS in the S30 system is highly active.

CFPS reactions have many components, and the cell lysate is the most complex. It is estimated that 500–850 proteins [36, 37] are present in the lysate, many of

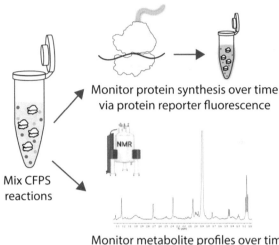

Fig. 2.3 Workflow for monitoring CFPS activity and metabolite profiles with NMR

Table 2.1 Components used in CFPS reactions

Reagent	[Final]	Reagent	[Final]
Magnesium glutamate	8 mM	CoA	0.26 mM
Ammonium glutamate	10 mM	Oxalic acid	4 mM
Potassium glutamate	130 mM	Putrescine	1 mM
ATP	1.2 mM	Spermidine	1.5 mM
GTP	0.85 mM	HEPES, pH 7.4	57 mM
UTP	0.85 mM	20 amino acids	2 mM
CTP	0.85 mM	PEP	30 mM
Folinic acid	0.034 mg/mL	Plasmid template	Varies
E. coli tRNA	0.17 mg/mL	S30 extract[a]	30% v/v
NAD	0.4 mM	Water	To volume

[a]S30 extract is referred to as lysate throughout the chapter

which are not necessary for the synthesis of the desired product. The composition of the lysate can fluctuate depending on the metabolic profiles of cells harvested, which change as a function of the stage of cell growth at the time of harvest. In addition to the cell lysate, CFPS reactions require a variety of reactants including amino acids and energy sources. Typical reagents for CFPS are shown in Table 2.1 [18, 38].

Many of these reactants have been studied in detail [20] to determine their optimal concentrations and if their inclusion in reactions is necessary. In this research, we are utilizing BL21 Star™ (DE3) as our bacterial chassis and a protocol that was optimized for phosphoenolpyruvate (PEP) as the initial energy source [18, 38, 39]. PEP is a high-energy compound derived from glycolysis and is highly regulated in

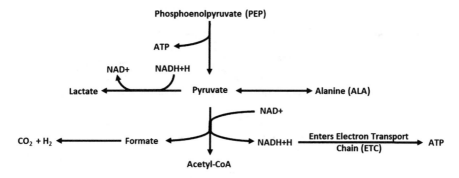

Fig. 2.4 Phosphoenolpyruvate catabolic pathways

biological systems. When one molecule of PEP is catabolized to pyruvate (PYR), one molecule of adenosine triphosphate (ATP) is formed via substrate-level phosphorylation. This first step produces a much-needed burst of energy to power translation and transcription. It is critical that ATP levels are sustained in the reaction. If they fall too low, product yield will suffer. Figure 2.4 depicts the initial reaction sequence and potential side paths at this stage of PEP metabolism [2].

The initial goal for the catabolism of PEP is to enter the tricarboxylic acid (TCA) cycle. Side paths might be activated, but higher yields are expected if most of the PEP eventually enters the TCA cycle via acetyl-CoA. This leads to increased ATP production that can occur during the TCA cycle and activation of the electron transport chain (ETC). The TCA cycle enables the lysate to produce more ATP than can be accomplished by substrate-level phosphorylation alone. Pushing the carbon resources toward the TCA cycle will result in sustained ATP production. Figure 2.5 illustrates the next stage of metabolism [2].

As PEP is catabolized, ATP is formed by substrate-level phosphorylation, and nicotinamide adenine dinucleotide (NADH) is produced in the TCA cycle. The NADH molecules will be converted back to NAD+ in the ETC, and this reaction will result in the production of three ATP molecules per NADH molecule via oxidative phosphorylation. If NADH cannot enter the ETC chain, the reaction will not produce enough ATP to fuel protein production. Ideally, PEP would be catabolized and enter the TCA cycle instead of producing metabolites such as lactate and ethanol. In this study, we focused on the metabolic profiles of the lysates and examined how they performed in reactions. Two lysates were produced in-house, RXAS 1 and RXAS 2, and two lysates were produced at Northwestern University in the Jewett Lab for cell-free synthetic biology, NU 1 and NU 2. All lysates were prepared utilizing the same bacterial chassis and protocol.

Each lysate was diluted with distilled water and monitored in the NMR for 3 hours. The purpose of this initial experiment was to determine if the lysates were metabolically active. No other reactants were added to the NMR tube to initiate CFPS. A comparison of all the lysates at time zero indicates that there are significant differences in the lysates as reported in Fig. 2.6a; peak identifications are

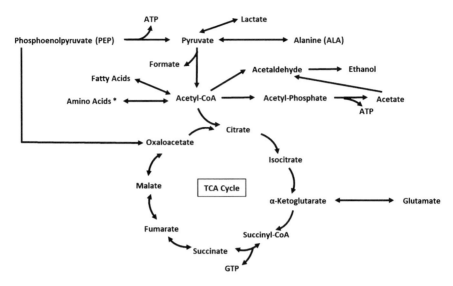

Fig. 2.5 Central metabolism in cell lysates

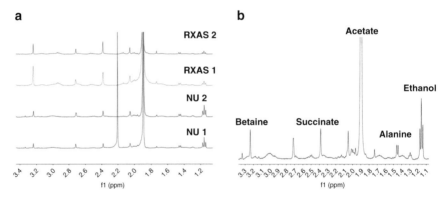

Fig. 2.6 Comparison of all lysates and peak identifications

depicted in Fig. 2.6b. The data in Fig. 2.6a for the diluted lysates illustrate the significant differences in the metabolic profiles; the metabolite peak intensities are shown in Table 2.2.

The peak intensities in Table 2.2 are normalized to the intensity of 4,4-dimethyl-4-silapentane-1-sulfonic acid (DSS), an internal NMR reference standard. Each intensity was also divided by the number of protons represented in the peak to enable 1:1 comparison. The peak data for the lysates indicate that all the lysates produce ethanol, alanine, and succinate over the 3-hour experiment. Even without exogenous reactants and a DNA template, the lysate is metabolically active and will conduct reactions utilizing the enzymes and substrates present in the lysate if conditions are

Table 2.2 Metabolite peak intensities for diluted lysates

	RXAS 1			RXAS 2			NU 1			NU 2		
	Initial	Final	Fold change	Initial	Final	Fold change	Initial	Final	Fold change	Initial	Final	Fold change
Alanine	0.80	1.16	1.45	0.57	0.81	1.43	0.90	1.04	1.16	0.94	1.18	1.25
Ethanol	1.48	3.61	2.44	1.06	3.38	3.18	3.89	5.33	1.37	4.16	6.56	1.58
Acetate	73.99	73.34	0.99	93.62	86.91	0.93	52.63	51.33	0.98	52.00	49.87	0.96
Succinate	1.74	2.28	1.31	1.61	2.29	1.43	0.65	0.68	1.05	0.76	0.92	1.22
Betaine	1.13	1.13	1.00	0.59	0.55	0.94	0.29	0.27	0.93	0.32	0.31	0.96

kinetically favorable. The high amounts of acetate in the lysates can be attributed to the buffer solutions utilized during lysate preparation as well as endogenous production of this metabolite. Acetate can be recycled back to acetyl-CoA at the cost of ATP. It was expected that the increase in ethanol would likely correlate with the decrease in acetate, but this did not occur. There are limited pathways that result in the production of ethanol, especially in the absence of an energy source such as PEP or pyruvate. The amino acids that could be catabolized to acetyl-CoA are below the level of detection in the cell lysates. It is possible that fatty acids are undergoing β-oxidation to yield acetyl-CoA [2]. When lysates are produced, any metabolites present at the time the living bacteria are lysed would be present in the lysate. Some metabolites, such as pyruvate, are short-lived and are unexpected in a prepared lysate unless an energy source is provided. The metabolites in the lysate are also driven by the environment. Anaerobic conditions will activate pathways that may not be active under aerobic conditions.

Another method to analyze the peak data is to use principal component analysis (PCA) [40]. PCA can utilize the entire NMR spectrum, unlike a few peaks as in the targeted analysis above, to determine the similarities between the lysates. The NMR spectra are binned prior to analysis, with each bin becoming a principal component. PCA is an unsupervised multivariate analysis technique. Figure 2.7 displays the PCA analysis of the lysates. The PCA score plot, Fig. 2.7a, illustrates that the four lysates are different, but the lysates from Northwestern University are the most similar to each other. RXAS 1 differs from the other three lysates along PC 1, and RXAS 2 differs mostly along PC 2. The difference between the RXAS lysates is interesting, as it appears that RXAS 2 may be more similar to the Northwestern lysates than RXAS 1. There is still room for improvement, as RXAS 2 does differ significantly along PC 2.

The principal component plot, Fig. 2.7b, reveals that almost 50% of the variance in the lysates is from PC 1 and that three principal components account for virtually all the variance between the lysates. To identify which metabolites are responsible for the variance, a discriminant analysis will need to be performed, such as OPLS-DA. Further statistical analysis is currently in progress to identify the metabolites responsible for the variance.

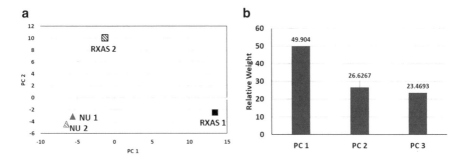

Fig. 2.7 PCA score plot and principal component plot

Table 2.3 Product yields of
CFPS reactions

Lysate	Product yield
RXAS 1	7 μM
RXAS 2	10.5 μM
NU 1	13 μM
NU 2	12.5 μM

After the lysate variabilities were assessed, we next examined if the lysates had significant differences in product yield. The CFPS reactions were assembled in an NMR tube, with all reactants except for the DNA template. The first NMR spectrum collected was the zero time for the reaction. The DNA was added to the NMR tube and subsequently analyzed for the next 3 hours. The reaction yields for 5 μL reactions, measured by plate reader in tandem with NMR experiments, are compared in Table 2.3.

It is evident that NU 1 achieves a much higher yield than RXAS 1. When examining the NMR data for RXAS 1, we found that the peaks of interest differ from the peaks of interest that were identified in the lysate analysis. The diluted lysates did not have detectable PEP peaks or any other peaks above 5 ppm. Once the lysate is supplied with reactants, the spectra become more complicated above 5 ppm. Of particular interest are the PEP peaks between 5.1 and 5.35 ppm, which are the primary energy sources for this reaction. Within 20 minutes, the PEP peak is below the limit of detection in reactions with the RXAS 1 lysate. The rapid loss of PEP in RXAS 1 reactions is illustrated in Fig. 2.8 as well as comparative plate reader data. Figure 2.8a displays the NMR data, and Fig. 2.8b provides yield data for 50 μL scale reactions. The catabolism of PEP is expected, but not this early into the reaction. The catabolism of PEP to pyruvate is an allosterically regulated process due to the cleavage of a high-energy phosphate group from PEP for substrate-level phosphorylation. If ATP is in excess, the catabolism of PEP will be reduced as ATP binds an inhibitory site on pyruvate kinase, the enzyme that performs the reaction. This rapid catabolism of PEP indicates that ATP must be low, as high ATP will inhibit the catabolism of PEP to pyruvate, which would allow the PEP to linger in solution until ATP levels fall. This finding may point to an energy crisis occurring in the RXAS 1 lysate. High levels of ATP are necessary to fuel mRNA production from the DNA template and protein production, and we can infer from these data that the low ATP production will ultimately lead to lower product yield.

The NU 1 reaction experiences a slower catabolism of PEP over the course of the reaction. Figure 2.9 displays the NMR data for a CFPS reaction with NU 1 lysate. NU 1 experiences a drop in PEP within the early stages of the reaction but maintains the PEP concentration into the next morning, roughly 15 hours after the experiment began. Pyruvate kinase is responsible for catalyzing the catabolism of PEP to pyruvate, which is inhibited by ATP, alanine, and citrate. The NU 1 reaction in Fig. 2.9, after the initial PEP catabolism, maintains the PEP concentration throughout the 3-hour reaction time, unlike the RXAS 1 reaction, which exhausted the PEP within 20 minutes. These data indicate that the NU 1 lysate is more efficient at producing ATP and must have a larger pool of ATP present in the reaction than the RXAS 1 lysate.

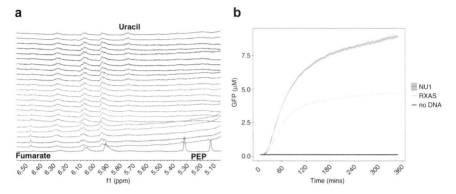

Fig. 2.8 NMR array and yield data for RXAS 1 reaction

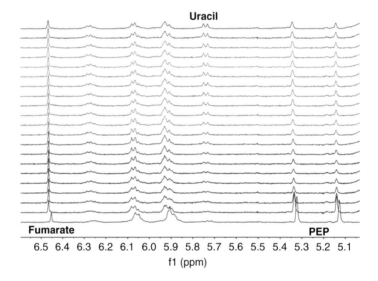

Fig. 2.9 NMR array for NU 1 reaction

Another metabolite of interest in the CFPS reactions is ethanol. Ethanol increases over the course of the CFPS reactions as shown in Table 2.4. The initial column indicates the first time point, and the final column is the last spectrum of the 3-hour experiment.

Comparing the RXAS 1 reaction and the NU 1 reaction, we observe a 51.15-fold change in ethanol production in the RXAS 1 reaction versus a 1.60-fold change in the NU 1 reaction. This implies that the NU 1 is producing less waste from the carbon resources compared to RXAS 1. It is interesting to note that, even if all the PEP in the RXAS 1 lysate during the reaction was eventually converted to ethanol, the production of ethanol far exceeds the amount of PEP provided. As discussed earlier, fatty acids present in the lysate could be the precursors to acetyl-CoA which could feed ethanol production. Fatty acids can undergo β-oxidation to yield acetyl-CoA,

Table 2.4 Metabolite peak intensities for CFPS reactions

	RXAS 1			NU 1		
	Initial	Final	*Fold change*	Initial	Final	*Fold change*
Alanine	14.52	14.98	*1.03*	14.35	12.83	*0.89*
Ethanol	13.30	680.20	*51.15*	7.17	11.45	*1.60*
Acetate	124.80	134.76	*1.08*	108.43	87.13	*0.80*
Glutamate	1206.14	690.41	*0.57*	1177.34	641.89	*0.55*
Succinate	11.55	21.00	*1.82*	13.87	21.67	*1.56*
PEP	6.54	0.00	*n/a*	6.35	0.41	*0.06*
Uracil	0.00	0.52	*n/a*	1.03	0.56	*0.76*
Fumarate	0.28	0.12	*0.42*	0.67	0.51	*0.54*
Formic acid	0.47	6.02	*12.71*	4.86	13.04	*2.68*

which is a direct precursor to ethanol. We are currently conducting experiments to verify if the fatty acid concentrations change during CFPS reactions and determine if the amount of change observed potentially accounts for the ethanol production.

2.3 Conclusions

NMR metabolomics has proven to be a fruitful approach to identifying metabolite profiles of the lysates and the CFPS reactions. The preliminary results indicate that ethanol may be a performance indicator of a poor-performing lysate. It is possible that inhibiting the production of ethanol in a lysate may affect the product yield, potentially improving the yield as carbon resources could be re-routed to produce more ATP for the reaction. The rate of PEP catabolism could also indicate the productivity of a given lysate; early depletion may indicate an energy crisis in the reaction. Future experiments will involve ^{13}C NMR to determine if fatty acid profiles are indeed changing. Also, ^{13}C-labeled PEP could be employed to track the metabolites of PEP created in the reaction which could identify pathway activations that are consuming carbon resources. In situ NMR metabolomics can provide metabolic guidance during reactions that could lead to yield improvements in the future.

References

1. E. Buchner, R. Rapp, Alkoholische Gährung ohne Hefezellen. Ber. Dtsch. Chem. Ges. **30**(3), 2668–2678 (1897). https://doi.org/10.1002/cber.18970300354
2. D.L. Nelson, A.L. Lehninger, M.M. Cox, *Lehninger Principles of Biochemistry* (W. H. Freeman, New York, 2008)
3. Eduard Buchner – Facts. NobelPrize.org. Nobel Media AB 2018 (2018), https://www.nobelprize.org/prizes/chemistry/1907/buchner/facts/. Accessed Wed. 5 Dec 2018
4. 2018 NMA (2018) Eduard Buchner – Nobel Lecture "Cell Free Fermentation". https://www.nobelprize.org/prizes/chemistry/1907/buchner/lecture/. Accessed Wed. 5 Dec 2018

5. J.K. Laylin, *Nobel Laureates in Chemistry, 1901–1992*, [Washington, D.C.]: American Chemical Society: Chemical Heritage Foundation, 1993
6. M.W. Nirenberg, J.H. Matthaei, The dependence of cell-free protein synthesis in E. coli upon naturally occurring or synthetic polyribonucleotides. Proc. Natl. Acad. Sci. U. S. A. **47**, 1588–1602 (1961)
7. P. Welch, R.K. Scopes, Studies on cell-free metabolism: Ethanol production by a yeast glycolytic system reconstituted from purified enzymes. J. Biotechnol. **2**(5), 257–273 (1985). https://doi.org/10.1016/0168-1656(85)90029-X
8. Y. Shimizu, A. Inoue, Y. Tomari, T. Suzuki, T. Yokogawa, K. Nishikawa, T. Ueda, Cell-free translation reconstituted with purified components. Nat. Biotechnol. **19**, 751 (2001). https://doi.org/10.1038/90802. https://www.nature.com/articles/nbt0801_751#supplementary-information
9. M.C. Jewett, K.A. Calhoun, A. Voloshin, J.J. Wuu, J.R. Swartz, An integrated cell-free metabolic platform for protein production and synthetic biology. Mol. Syst. Biol. **4**(1), 220 (2008). https://doi.org/10.1038/msb.2008.57
10. J.F. Zawada, G. Yin, A.R. Steiner, J. Yang, A. Naresh, S.M. Roy, D.S. Gold, H.G. Heinsohn, C.J. Murray, Microscale to manufacturing scale-up of cell-free cytokine production – A new approach for shortening protein production development timelines. Biotechnol. Bioeng. **108**(7), 1570–1578 (2011). https://doi.org/10.1002/bit.23103
11. K. Pardee, A.A. Green, M.K. Takahashi, D. Braff, G. Lambert, J.W. Lee, T. Ferrante, D. Ma, N. Donghia, M. Fan, N.M. Daringer, I. Bosch, D.M. Dudley, D.H. O'Connor, L. Gehrke, J.J. Collins, Rapid, low-cost detection of Zika virus using programmable biomolecular components. Cell **165**(5), 1255–1266 (2016). https://doi.org/10.1016/j.cell.2016.04.059
12. C.E. Nakamura, G.M. Whited, Metabolic engineering for the microbial production of 1,3-propanediol. Curr. Opin. Biotechnol. **14**(5), 454–459 (2003). https://doi.org/10.1016/j.copbio.2003.08.005
13. Military Specification MIL-PRF-5606J, Hydraulic Fluid, Petroleum base; Aircraft, Missile, and ordnance. (DOD, 5 March 2018) (2018). https://assist.dla.mil/online/start/
14. Military Specification MIL-DTL-83133. Turbine Fuel, Aviation, Kerosene Type, JP-8 (NATO F-34), NATO F-35, and JP-8+100 (NATO F-37). DOD 18 July 2018
15. Energy USDo (2019) Biomass basics: The facts about bioenergy. U.S. Department of Energy. Accessed 1 May 2019
16. Q.M. Dudley, K.C. Anderson, M.C. Jewett, Cell-free mixing of Escherichia coli crude extracts to prototype and rationally engineer high-titer mevalonate synthesis. ACS Synth. Biol. **5**(12), 1578–1588 (2016). https://doi.org/10.1021/acssynbio.6b00154
17. J.E. Kay, M.C. Jewett, Lysate of engineered Escherichia coli supports high-level conversion of glucose to 2,3-butanediol. Metab. Eng. **32**, 133–142 (2015). https://doi.org/10.1016/j.ymben.2015.09.015
18. A.S. Karim, M.C. Jewett, A cell-free framework for rapid biosynthetic pathway prototyping and enzyme discovery. Metab. Eng. **36**, 116–126 (2016). https://doi.org/10.1016/j.ymben.2016.03.002
19. E.D. Carlson, R. Gan, C.E. Hodgman, M.C. Jewett, Cell-free protein synthesis: Applications come of age. Biotechnol. Adv. **30**(5), 1185–1194 (2012). https://doi.org/10.1016/j.biotechadv.2011.09.016
20. A.S. Karim, J.T. Heggestad, S.A. Crowe, M.C. Jewett, Controlling cell-free metabolism through physiochemical perturbations. Metab. Eng. **45**, 86–94 (2018). https://doi.org/10.1016/j.ymben.2017.11.005
21. J. Li, H. Wang, Y.-C. Kwon, M.C. Jewett, Establishing a high yielding streptomyces-based cell-free protein synthesis system. Biotechnol. Bioeng. **114**(6), 1343–1353 (2017). https://doi.org/10.1002/bit.26253
22. M. Buntru, S. Vogel, K. Stoff, H. Spiegel, S. Schillberg, A versatile coupled cell-free transcription–translation system based on tobacco BY-2 cell lysates. Biotechnol. Bioeng. **112**(5), 867–878 (2015). https://doi.org/10.1002/bit.25502

23. M.J. Anderson, J.C. Stark, C.E. Hodgman, M.C. Jewett, Energizing eukaryotic cell-free protein synthesis with glucose metabolism. FEBS Lett. **589**(15), 1723–1727 (2015). https://doi.org/10.1016/j.febslet.2015.05.045
24. W.R.S. Trindade, R.G. Santos, Review on the characteristics of butanol, its production and use as fuel in internal combustion engines. Renew. Sust. Energ. Rev. **69**, 642–651 (2017). https://doi.org/10.1016/j.rser.2016.11.213
25. L. Siwale, L. Kristóf, T. Adam, A. Bereczky, M. Mbarawa, A. Penninger, A. Kolesnikov, Combustion and emission characteristics of n-butanol/diesel fuel blend in a turbo-charged compression ignition engine. Fuel **107**, 409–418 (2013). https://doi.org/10.1016/j.fuel.2012.11.083
26. Y. Li, W. Tang, Y. Chen, J. Liu, C.-fF. Lee, Potential of acetone-butanol-ethanol (ABE) as a biofuel. Fuel **242**, 673–686 (2019). https://doi.org/10.1016/j.fuel.2019.01.063
27. M. Yao, H. Wang, Z. Zheng, Y. Yue, Experimental study of n-butanol additive and multi-injection on HD diesel engine performance and emissions. Fuel **89**(9), 2191–2201 (2010). https://doi.org/10.1016/j.fuel.2010.04.008
28. N. Vivek, L.M. Nair, B. Mohan, S.C. Nair, R. Sindhu, A. Pandey, N. Shurpali, P. Binod, Bio-butanol production from rice straw – Recent trends, possibilities, and challenges. Bioresource Technology Reports 7, 100224 (2019). https://doi.org/10.1016/j.biteb.2019.100224
29. D. Eady, S.B. Siegel, R.S. Bell, S.H. Dicke, Sustain the mission project: Casualty factors for fuel and water resupply convoys (2009)
30. D.S. Board, *Task Force on Energy Systems for Forward/Remote Operating Bases* (Office of the Under Secretary of Defense for Acquisition, Technology, and Logistics, Washington D.C., 2016)
31. P. Godoy, Á. Mourenza, S. Hernández-Romero, J. González-López, M. Manzanera, Microbial production of ethanol from sludge derived from an urban wastewater treatment plant. Front. Microbiol. **9**, 2634–2634 (2018). https://doi.org/10.3389/fmicb.2018.02634
32. Defense Do, Exposure to toxins produced by burn pits: Congressional data request and studies, in Memorandum for the Assistant Secretary of Defense for Health Affairs, Washington, D.C. (2011)
33. T. Dominguez, J. Aurell, B. Gullett, R. Eninger, D. Yamamoto, Characterizing emissions from open burning of military food waste and ration packaging compositions. **20** (2017). https://doi.org/10.1007/s10163-017-0652-y
34. J.F. Sopko, Final assessment: What we have learned from our inspections of incinerators and use of burn pits in Afghanistan. Special Inspector General for Afghanistan Reconstruction, Arlington (February 2015)
35. K. Rock, L. Lesher, M.F. Kramer, J. Johnson, M. Bordic, H. Miler, An analysis of military field-feeding waste (2000) [Natick MA, United States]: Army Soldier and Biological Chemical Command Soldier Systems Center
36. D. Garenne, C.L. Beisel, V. Noireaux, Characterization of the all-E. coli transcription-translation system myTXTL by mass spectrometry. Rapid Commun. Mass Spectrom. **33**(11), 1036–1048 (2019). https://doi.org/10.1002/rcm.8438
37. D. Foshag, E. Henrich, E. Hiller, M. Schäfer, C. Kerger, A. Burger-Kentischer, I. Diaz-Moreno, S.M. García-Mauriño, V. Dötsch, S. Rupp, F. Bernhard, The E. coli S30 lysate proteome: A prototype for cell-free protein production. New Biotechnol. **40**, 245–260 (2018). https://doi.org/10.1016/j.nbt.2017.09.005
38. M.C. Jewett, J.R. Swartz, Substrate replenishment extends protein synthesis with an in vitro translation system designed to mimic the cytoplasm. Biotechnol. Bioeng. **87**(4), 465–471 (2004). https://doi.org/10.1002/bit.20139
39. Y.-C. Kwon, M.C. Jewett, High-throughput preparation methods of crude extract for robust cell-free protein synthesis. Sci. Rep. **5**, 8663 (2015). https://doi.org/10.1038/srep08663. https://www.nature.com/articles/srep08663#supplementary-information
40. J. Bartel, J. Krumsiek, F.J. Theis, Statistical methods for the analysis of high-throughput metabolomics data. Comput. Struct. Biotechnol. J. **4**(5), e201301009 (2013). https://doi.org/10.5936/csbj.201301009

Chapter 3
Development of Organic Nonlinear Optical Materials for Light Manipulation

Joy E. Haley

3.1 Introduction

The development of nonlinear optical materials for light manipulation has been an area of interest for the US Air Force for over three decades. In particular an interest in organic or organometallic materials has persisted and led to the development of materials with larger nonlinearities based on design. Application of these materials includes optical data storage [1], frequency-upconverted lasing [2], nonlinear photonics [3], microfabrication [4], fluorescence imaging [5], and photodynamic therapy [6]. Chemical synthesis provides the flexibility to make many variations on material structure to investigate structure-property relationships. This has led to a rich history in the development and deep understanding of optical properties of several classes of materials including platinum acetylides, AFX dyes, combined platinum acetylide with AFX dyes, porphyrins, and phthalocyanines.

The nonlinear mechanisms that are relevant include nonlinear absorption (the imaginary part of χ^3), nonlinear refraction (the real part of χ^3), and stimulated scattering [7]. In general, the organic materials are nonlinear absorbers, so we have focused our efforts on understanding the properties that contribute to nonlinear absorption. Figure 3.1 shows an energy-level diagram known as a Jablonski diagram that describes the fate of an electron upon absorption of light. Initially the electron is promoted to an upper level singlet excited state (S_n). The electron then quickly falls to the S_1 level as defined by Kasha's rule [8]. From there the fate of the electron is determined by the inherent properties of the material. There are three typical pathways ($k_{S1} = k_f + k_{IC} + k_{ISC}$). The first is that the electron will decay back to the ground state and give off light in the form of fluorescence (k_f). The second pathway is that the electron will return to the ground state and give off heat in the

J. E. Haley (✉)
Materials and Manufacturing Directorate, Air Force Research Laboratory,
Wright-Patterson Air Force Base, OH, USA
e-mail: joy.haley.1@us.af.mil

© Springer Nature Switzerland AG 2020   35
M. E. Kinsella (ed.), *Women in Aerospace Materials*, Women in Engineering
and Science, https://doi.org/10.1007/978-3-030-40779-7_3

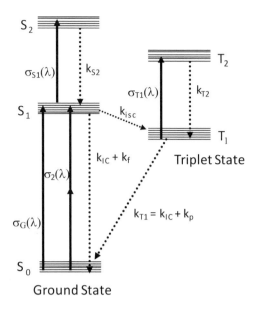

Fig. 3.1 Energy-level diagram defining promotion of an electron upon light absorption

form of internal conversion (k_{IC}). The last pathway is that the electron will undergo a spin flip to form a triplet excited state through intersystem crossing (k_{ISC}). Once the electron is in the triplet excited state, it will decay through two mechanisms. It will either give off light in the form of phosphorescence (k_p) or it will give off heat in the form of internal conversion (k_{IC}). k_{T1}, the overall decay from the triplet excited state, is defined by $k_{T1} = k_{IC} + k_p$. The interesting part of the triplet excited state is that it takes some time for the electron to flip its spin back to return to the original singlet-derived electron pair. Therefore, the lifetime of the triplet excited state is longer, typically on the order of microseconds.

It is known that the singlet and triplet excited-state absorption of an organic or organometallic molecule contributes to the nonlinear absorption through two distinct mechanisms. The first mechanism is known as reverse saturable absorption and is defined as an area where the singlet excited-state absorption (σ_{S1}) or the triplet excited-state absorption (σ_{T1}) is larger than the ground-state absorption (σ_G) as defined in Fig. 3.1. The second mechanism is a two-photon-assisted excited-state absorption. This is described in Fig. 3.1 where two photons of lower energy are absorbed simultaneously (σ_2) to get to the singlet excited state followed by a sequential absorption of a photon into either the singlet excited state or the triplet excited state. There is a third mechanism of just two-photon absorption (σ_2) that occurs under femtosecond excitation, but this will not be discussed further herein.

Typically, third-order nonlinear properties are measured for a material utilizing techniques such as Z-scan or I-scan where the material is evaluated under different intensities of light at various pulse widths from femtosecond to nanosecond. While these measurements are an important part of evaluating a material, there are several factors that may be independently measured that contribute to the nonlinearity of a material and allow for modeling of the data [9]. These include measurements such as two-photon cross-section values (σ_2), ground-state cross-section values (σ_G), singlet or triplet excited-state cross-section values (σ_{S1}, σ_{T1}), intersystem crossing quantum yields (Φ_{ISC}), and excited-state lifetimes (τ_{S1}, τ_{T1}). Currently within the Air Force Research Laboratory, we have developed techniques to measure all of these properties independently. Through the years, we have reported on these measurements in the literature [10, 11]. Briefly we utilize ultraviolet/visible (UV/Vis) absorbance techniques to measure ground-state cross sections; both steady-state and time-resolved emission for lifetimes and knowledge of fluorescence, internal conversion, and intersystem crossing; and both femtosecond and nanosecond transient absorption techniques to measure singlet and triplet cross sections and intersystem crossing quantum yields. Lastly, we utilize femtosecond Z-scan or femtosecond fluorescence techniques to measure two-photon cross-section values when necessary.

Overall understanding of structure-property relationships within a series of materials has informed the group on the best nonlinear optical materials for a given application. Through the years, we have looked at hundreds of new light-absorbing materials, or chromophores, but within this article, we will only touch on four classes of significance.

3.2 Overview of Classes of Nonlinear Optical Materials

Platinum Acetylides Shown in Fig. 3.2 are the structures of a series of platinum acetylides synthesized and studied at AFRL [11]. These materials are reverse saturable absorbers, meaning that their excited-state absorption is larger than their ground-state absorption, with no two-photon activity. In this study the effect of ligand length on both ground and triplet excited properties was determined. Shown in Fig. 3.3a are the T_1-T_n absorption spectra of the PPE, PE2, and PE3 upon 355 nm excitation. All show broad T_1-T_n absorption from 400 to 700 nm. Figure 3.3b shows the single exponential decays of the triplet excited state under deoxygenated conditions. Overall it was learned through this structure-property study that by extending the π-conjugated ligand, an overall red shift in the ground-state and excited-state absorption and an increase in the molar absorption coefficient are observed. As the conjugation length increases, an increase is observed in the ΔE_{ST}, defined as the energy difference between the singlet and triplet excited state, and results in a believed slower rate for intersystem crossing, thus leading to a smaller triplet quantum yield. Also with increased conjugation, the triplet excited-state lifetime becomes longer, which is proposed to be due to the spin-orbit coupling effect being reduced.

AFX Dyes For many years now, AFRL has been a leader in the synthesis of small organic two-photon-absorbing chromophores. This work began in the early 1990s and still remains relevant today [12]. Shown in Fig. 3.4 is the generation of some of the AFX two-photon chromophores through the years as a measure of "effective"

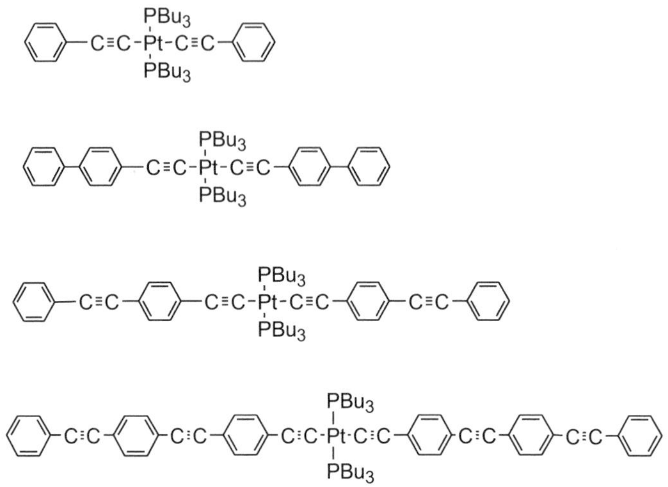

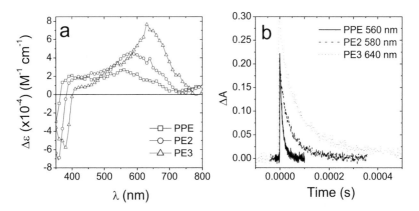

Fig. 3.2 Structures of platinum acetylides – PE1, PPE, PE2, and PE3. (Reprinted with permission from Rogers [11]. Copyright 2002 American Chemical Society)

Fig. 3.3 T_1-T_n absorption spectra observed after nanosecond pulsed 355 nm excitation of PPE (12.9 μM), PE2 (4.5 μM), and PE3 (5.2 μM) in argon-saturated benzene. Molar absorption coefficients were obtained using the methods of triplet sensitization and singlet depletion as described in the text. *Panel b:* Shown are the single exponential decays of the triplet excited state obtained after three freeze-pump-thaw cycles to remove oxygen. (Reprinted with permission from Rogers [11]. Copyright 2002 American Chemical Society)

two-photon cross-section values given in units of GM measured at 800 nm with 5 ns pulses. The measurements were made in Paras Prasad's lab at the University of Buffalo. Structure-property studies afford us the understanding of how small changes to the molecule affect the overall nonlinearity in these materials. Interestingly the intrinsic two-photon cross-section values are only a small part of the puzzle. It was found through singlet and triplet excited-state characterization

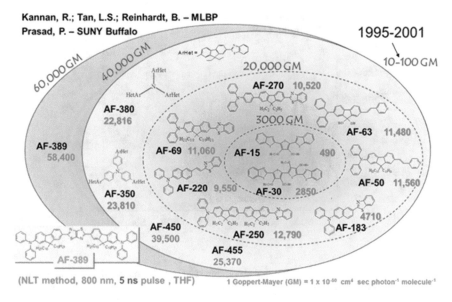

Fig. 3.4 Evolution of design in AFX two-photon-absorbing materials. Values shown are effective two-photon cross sections measured at 800 nm with 5 ns pulses in air-saturated THF. Measurements were made by the group of Paras Prasad at University of Buffalo

that these materials have larger "effective" two-photon cross-section values at 800 nm that directly correlate to large singlet and triplet excited-state absorption [13]. For example, shown in Fig. 3.5 are the S_0-S_1, S_1-S_n, and T_1-T_n absorption and two-photon absorption for AF455 in deoxygenated tetrahydrofuran (THF) [14]. A key point to observe is that the large triplet state absorption of AF455 is a major contributor to the "effective" two-photon cross-section value of 25,370 GM at 800 nm due to the large peak centered around 900 nm. In addition, the two-photon absorption is also in the same region. Through continued work in this area, it was found that one of the main contributors to overall "effective" nonlinearity under nanosecond pulses is the ability of the AFX dye to form a triplet excited state and have triplet excited-state absorption at 800 nm. In general, these materials do not go effectively to the triplet excited state and have small triplet quantum yields of less than 10%.

Efforts have been made to try to increase the triplet quantum yield and the intrinsic two-photon cross section through structure variation. A recent study looked at the effect of steric hindrance on the formation of an intramolecular charge transfer (ICT) state and how this correlates to the strength of the two-photon cross section [15]. The structures of the AF240 derivatives are shown in Fig. 3.6. It was found that compounds 2 and 3, containing sterically hindering alkyl groups, show less ICT character than compound 1 despite the electron-donating ability of those groups. When the bulky groups are placed in the para position of the donor moiety in compound 4, thus eliminating the steric crowding, a clear increase in ICT character is

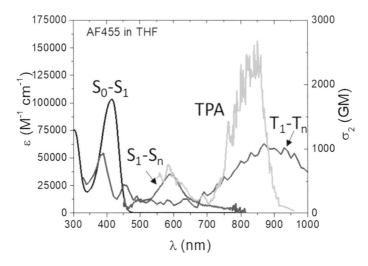

Fig. 3.5 S_0-S_1, S_1-S_n, and T_1-T_n quantified absorption spectra of AF455 in THF. Also shown on the right axis are the two photon cross-section values for AF455 in THF

Fig. 3.6 Structures of sterically hindered AF240 dyes. (Stewart [15]. Reproduced by permission of the PCCP Owner Societies)

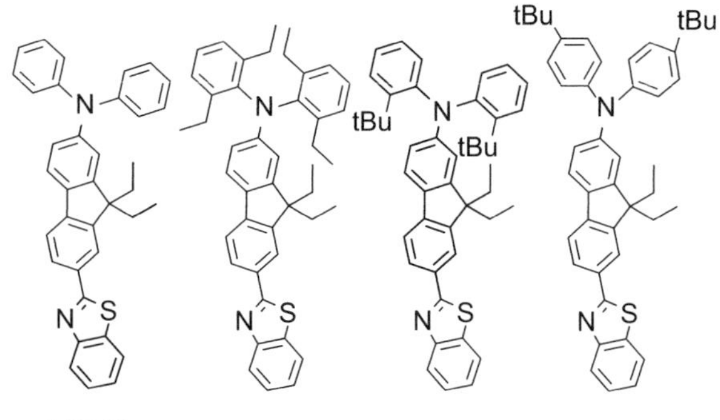

observed with respect to compound 1. This supports the notion that the steric crowding inhibits geometric relaxation after photoexcitation and is consistent with a PICT (planar intramolecular charge transfer) excited state. This study is just one example of looking very carefully at structure-property relationships with the AFX compounds. The overall results from these studies show that small structural changes do affect not only the intrinsic two-photon properties but also the singlet and triplet excited-state properties that correlate to the overall nonlinear optical properties.

Combined Platinum Acetylide/AFX Dyes One key point from the platinum acetylide study was that these materials only behave as reverse saturable absorbers from roughly 400 to 550 nm, which is a very short range. So, in an effort to increase their nonlinear behavior wavelength range, a new class of materials was developed

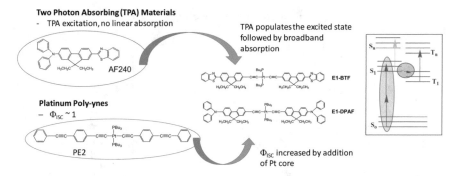

Fig. 3.7 Approach of marrying the two-photon-absorbing materials with the platinum poly-ynes that leads to broad nonlinear absorbing materials

that is both two-photon active and reverse saturable absorbers. Depicted in Fig. 3.7 is the approach of marrying a known two-photon absorbing (TPA) material with the platinum poly-ynes (aka acetylides) [10]. E1-BTF has electron-withdrawing ligands, while E1-DPAF has electron-donating ligands. Fortunately, the concept worked and these materials show both RSA and two-photon behavior, depending on the wavelength tested. Shown in Fig. 3.8 is the two-photon spectrum for E1-BTF (structure in Fig. 3.7) synthesized at the Air Force Research Laboratory. The absolute two-photon absorption cross sections were measured using the sample fluorescence relative to a bis-diphenylaminostilbene solution in methylene chloride. This method gives the cross-section value σ'_2, which, according to definition, is twice that which would be obtained with nonlinear absorption techniques ($\sigma'_2 = 2\,\sigma_2$) [16]. The design of new two-photon-absorbing chromophores coupled with Pt complexes to produce materials that exhibit large intrinsic two-photon-absorbing cross sections coupled with efficient intersystem crossing to afford long-lifetime triplet states has proven successful in this study. The conjugated ligands impart the complexes with effective two-photon absorption properties, while the heavy metal platinum centers give rise to efficient intersystem crossing to afford long-lived triplet states. This material also maintains its broad triplet excited-state absorption that is necessary for strong nonlinear absorbance.

Porphyrins and Phthalocyanines The last class of nonlinear absorbing materials that we have studied are the porphyrins and phthalocyanines. These materials are reverse saturable absorbers (RSA) meaning that their excited-state cross section is larger than their ground-state cross section. An example of an RSA material is given in Fig. 3.9. Reverse saturable absorption happens in the region where the blue arrow is defined where $\sigma_g < \sigma_T$. Here we have also focused our efforts on looking at structure-property relationships.

One of the first studies done by our lab was to look at the effect of annulation to a porphyrin by adding benzene rings to the pyrrole group in the form of benzoporphyrins and napthoporphyrins [17]. The structures are shown in Fig. 3.10 for this

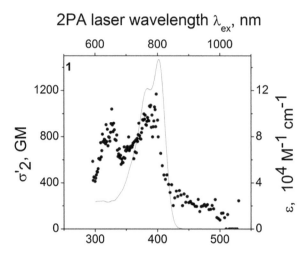

Fig. 3.8 Two-photon absorption spectrum of E1-BTF dissolved in benzene at ambient conditions (symbols). The samples were excited with 100 fs pulses with λ_{ex} 550–1100 nm. One-photon absorption spectra (solid lines) are also shown for comparison. Bottom x-axis presents transition wavelength (which is equal to 1PA wavelength), and top x-axis presents laser wavelength, used for 2PA excitation. (Reprinted with permission from Rogers et al. [10]. Copyright 2007 American Chemical Society)

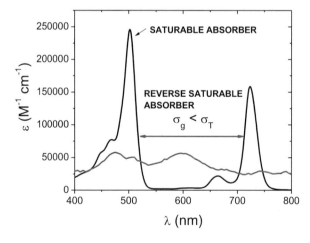

Fig. 3.9 Reverse saturable absorption shown for a porphyrin material. The area where $\sigma_g < \sigma_T$ is the region of interest

series where we also varied the metal by looking at both a zinc (Zn) series and a palladium (Pd) series. In this study we were interested in broadening the distance between the S_0-S_2 transition (Q-band) and the S_0-S_1 (B-band) transition. Extension of the porphyrin ring leads to dramatic red shifts in the ground-state spectra that are mainly due to increased conjugation effects with contributions from distortion. The

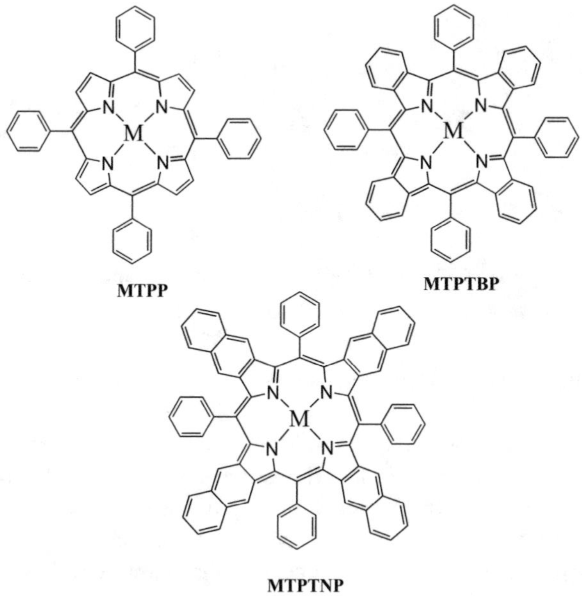

Fig. 3.10 Shown are the structures of the porphyrin series MTPP (meso-tetraphenylporphyrin), MTPTBP (meso-tetraphenylbenzoporphyrin), and MTPTNP (meso-tetraphenylnaphthoporphyrin). The metal (M) is either Pd or Zn. (Reprinted with permission from Rogers et al. [17]. Copyright 2003 American Chemical Society)

red shifts observed for these annulated porphyrins can be traced to the destabilization of the HOMO-1 (highest occupied molecular orbital) of the tetraphenylporphyrins. A red shift in the ground-state spectra was observed when changing the central metal from Pd to Zn, thus indicating involvement of the metal in the optical transitions. The kinetic parameters of these chromophores are more dependent on the central metal than on the extension of the π system. A significant heavy atom effect was observed in Pd porphyrins.

We have also investigated other series of porphyrins and phthalocyanines based on structure-property relationships and found that they too offer varied properties based on small changes to the structure [18–20].

3.3 Lessons Learned in Development and Utility for AF Applications

Initial studies on nonlinear optical absorbing materials were all done in a solution phase, but for real Air Force applications, there is a need to have these dyes embedded in a solid matrix. Engineering a liquid is much more difficult than dealing with a solid material. Therefore, we started to investigate several classes of host matrices.

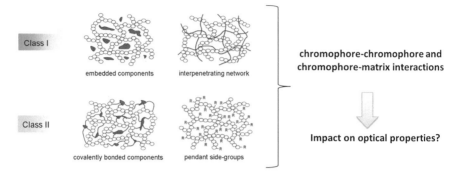

Fig. 3.11 Two types of hosts – Class I and Class II

Shown in Fig. 3.11 are the two types of hosts. Class I materials are simply embedded with the polymer, whereas a Class II material is actually covalently bound to the backbone of the polymer. Interestingly, putting a dye into a polymer host leads to new optical phenomena that need to be characterized. How do the chromophores interact when packed together at high concentration? How does the material interact with the host matrix? Ultimately how does this impact the optical properties?

One of the initial studies done in a polymer matrix was looking at E1-BTF (Fig. 3.12) in polymethylmethacrylate (PMMA) [21]. PMMA is a Class I host in which the E1-BTF is not covalently bound to the polymer. Results from this study reveal the formation of excimers (excited-state dimers) with an increase in concentration. Excimers form from the triplet excited state of the E1-BTF as determined through transient absorption techniques. Kinetically the formation is rather slow, but due to high concentration of the E1-BTF, the excimer is readily formed. This must be considered when making nonlinear absorption measurements since the excimer will certainly contribute to the overall nonlinearity. The excimer creates a separate kinetic pathway for the excited state to decay taking away from the needed triplet excited-state absorption in the nonlinear process.

More recently we have had success in getting multiple dyes into different host materials. Figure 3.13 shows both a two-photon dye, AF455, and a dual RSA – two-photon dye, E1-BTF. These materials have been placed in hosts such as urethane, epoxy, sol-gel, and PMMA. The urethane and sol-gel both represent Class II host materials, and the epoxy and PMMA are both Class I hosts. These materials are all optically clear and have been cut and polished for testing. In general the best success has been with cross-linked systems like urethane and sol-gel. Here concentrations approaching 0.5 M have been achieved. High concentrations are necessary in two-photon-absorbing materials because the two-photon absorption process is dependent on concentration of material.

Results show that the interaction of the chromophores under high packing has an adverse effect on the nonlinear performance of the material. This is mainly due to the formation of an excimer as observed in the PMMA data shown in Fig. 3.12. Instead of the triplet excited state being formed, the singlet excited state decays

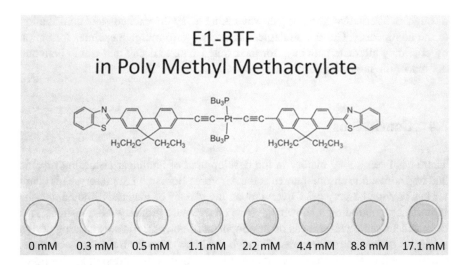

Fig. 3.12 Structure of the chromophore E1-BTF including pictures of the PMMA samples prepared. E1-BTF concentrations shown are 0 mM, 0.3 mM, 0.5 mM, 1.1 mM, 2.2 mM, 4.4 mM, 8.8 mM, and 17.1 mM

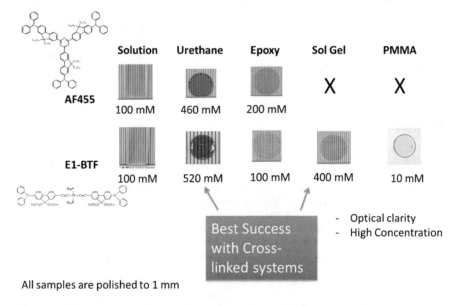

Fig. 3.13 AF455 and E1-BTF solid samples including urethane, epoxy, sol-gel, and PMMA hosts. All are optically clear and high concentrations have been achieved

through an alternative excimer pathway taking away the excited-state contribution to the nonlinearity. Current strategies include ways to mitigate excimer formation by providing alternate pathways for the excited state to decay that would help and not hinder the nonlinear absorption process.

3.4 Conclusions

There has been a long history in the development of nonlinear absorbing organic and organometallic chromophores at the Air Force Research Laboratory. This chapter has provided a very brief overview on the classes of materials studied and the importance of building a knowledge base on structure-property relationships. This deep understanding informs us on development of new materials by altering design strategies of new chromophores. Through the knowledge from this body of work, we will continue to develop the next generation of materials to support the Air Force mission.

Acknowledgments I would like to acknowledge the numerous contributors to this body of work including Douglas Krein, Thomas Cooper, Loon-Seng Tan, Jonathan Slagle, Daniel McLean, Weijie Su, David Stewart, Rammamurthi Kannan, Zhenning Yu, Tod Grusenmeyer, Aaron Burke, Paul Fleitz, Matthew Dalton, Kristi Singh, Augustine Urbas, Benjamin Hall, Paras Prasad, Aleks Rebane, and many students who have worked with us. I would also like to recognize the support of this work through both the Materials and Manufacturing Directorate and Air Force Office of Scientific Research of the Air Force Research Laboratory.

References

1. (a) D.A. Parthenopoulos, P.M. Rentzepis, Science. **249**, 843 (1989). (b) A.S. Dvornikov, P.M. Rentzepis. Opt. Commun. **119**, 341 (1995)
2. (a) J.D. Bhawalkar, G.S. He, P.N. Prasad, Rep. Prog. Phys. **59**, 1041 (1996); (b) G.S. He, C.F. Zhao, J.D. Bhawalkar, P.N. Prasad, Appl. Phys. Lett. **78**, 3703 (1995) (c) C.F. Zhao; G.S. He; J.D. Bhawalkar; C.K. Park; P.N. Prasad, Chem. Mater. **7**, 1979 (1995)
3. (a) P.A. Fleitz, R.A. Sutherland, F.P. Strogkendl, F.P. Larson, L.R. Dalton, SPIE Proc. **3472**, 91 (1998); (b) G.S. He, J.D. Bhawalkar, C.F. Zhao, P.N. Prasad, Appl. Phys. Lett. **67**, 2433 (1995); (c) J.E. Ehrlich, X.L. Wu, L.Y. Lee, Z.Y. Hu, H. Roeckel, S.R. Marder, J. Perry, Opt. Lett. **22**, 1843 (1997)
4. (a) S. Kawata, H.B. Sun, T. Tanaka, K. Takada, Nature **412**, 697–698 (2001); (b) B.H. Cumpston, S.P. Ananthavel, S. Barlow, D.L. Dyer, J.E. Ehrlich, L.L. Erskine, A.A. Heikal, S.M. Kuebler, I.Y.S. Le, D. McCord-Maughon, J. Qin, H. Rockel, M. Rumi, X.L. Wu, S.R. Marder, J.W. Perry, Nature, **398**, 51 (1999)
5. W. Denk, J.H. Strickler, W.W. Webb, Science **248**, 73 (1990)
6. J.D. Bhawalkar, N.D. Kumar, C.F. Zhao, P.N. Prasad, J. Clin. Laser Med. Surg. **15**, 201 (1997)
7. R.L. Sutherland, *Handbook of Nonlinear Optics* (Marcel-Dekker, Inc., New York, 2003)
8. M. Kasha, Disc. Faraday Soc. **9**, 14 (1950)
9. R.L. Sutherland, M.C. Brant, J. Heinrichs, J.E. Rogers, J.E. Slagle, D.G. McLean, P.A. Fleitz, J. Opt. Soc. Am. B **22**, 1939 (2005)

10. J.E. Rogers, J.E. Slagle, D.M. Krein, A.R. Burke, B.C. Hall, A. Fratini, D.G. McLean, P.A. Fleitz, T.M. Cooper, M. Drobizhev, N.S. Makarove, A. Rebane, K.-Y. Kim, R. Farley, K.S. Schanze, Platinum acetylide two-photon chromophores. Inorg. Chem. **46**, 6483 (2007)
11. J.E. Rogers, T.M. Cooper, P.A. Fleitz, D.J. Glass, D.G. McLean, J. Phys. Chem. A **106**, 10108–10115 (2002)
12. G.S. He, L.-S. Tan, Q. Zheng, P.N. Prasad, Chem. Rev. **108**, 1245–1330 (2008)
13. R.L. Sutherland, M.C. Brant, J. Heinrichs, J.E. Rogers, J.E. Slagle, D.G. McLean, P.A.J. Fleitz, Opt. Sci. Am. B **22**, 1939–1948 (2005)
14. J.E. Rogers, J.E. Slagle, D.G. McLean, R.L. Sutherland, B. Sankaran, R. Kannan, L.-S. Tan, P.A. Fleitz, J. Phys. Chem. A **108**, 5514–5520 (2004)
15. D.J. Stewart, M.J. Dalton, S.L. Long, R. Kannan, Z. Yu, T.M. Cooper, J.E. Haley, L.-S. Tan, Phys. Chem. Chem. Phys. **18**, 5587–5596 (2016)
16. M. Drobizhev, Y. Stepanenko, Y. Dzenis, A. Karotki, A. Rebane, P.N. Taylor, H.L. Anderson, J. Phys. Chem. B **109**, 7223 (2005)
17. J.E. Rogers, K.A. Nguyen, D.C. Hufnagle, D.G. McLean, W. Su, K.M. Gossett, A.R. Burke, S.A. Vinogradov, R. Pachter, P.A.J. Fleitz, Phys. Chem. A **107**, 11331–11339 (2003)
18. K.J. McEwan, P.A. Fleitz, J.E. Rogers, J.E. Slagle, D.G. McLean, H. Akdas, M. Katterle, I.M. Blake, H.L. Anderson, Adv. Mater. **16**, 1933–1935 (2004)
19. M.J. Frampton, G. Accorsi, N. Araroli, J.E. Rogers, P.A. Fleitz, K.J. McEwan, H.L. Anderson, Org. Biomol. Chem. **5**, 1056–1061 (2007)
20. J.E. Haley, W. Su, K.M. Singh, J.L. Monahan, J.E. Slagle, D.G. Mclean, T.M.J. Cooper, Porphyrins Phthalocyanines **15**, 1–10 (2011)
21. J.E. Haley, J.L. Flikkema, W. Su, D.M. Krein, T.M. Cooper, MRS Symp. Proc., *MRSF11-1392-K04-02.R1*, (2012)

Chapter 4
2D Materials: Molybdenum Disulfide for Electronic and Optoelectronic Devices

Shanee Pacley

4.1 Introduction

Two-dimensional (2D) materials such as graphene, molybdenum disulfide (MoS_2), tungsten diselenide (WSe_2), black phosphorus, and boron nitride (BN) have attracted much attention due to their extraordinary electronic and optical properties, making them ideal candidates for next-generation electronic and optoelectronic devices [1–4]. In particular, a monolayer of MoS_2 has a direct bandgap of 1.8–1.9 eV [5, 6], making it an ideal candidate for the mentioned applications [1, 2]. Growth processes of 2D MoS_2 include mechanical exfoliation [7–9], chemical vapor deposition (CVD) [5, 6], intercalation-assisted exfoliation [10–13], physical vapor deposition [14, 15], metal organic chemical vapor deposition [16], and a wet chemistry approach involving thermal decomposition of a precursor containing Mo and S [17]. An advantage of CVD growth of MoS_2 is the ability to grow large area films for device fabrication. Molybdenum disulfide films grown using CVD have demonstrated promising results for semiconductor grade material properties, with observed field-effect mobilities around 500 cm^2/Vs [18]. During CVD growth, sulfurization of molybdenum-containing precursors such as Mo, MoO_3, and $MoCl_5$ is usually performed. In the case of MoO_3 [6] and $MoCl_5$ [19], the precursors have been powders or ribbons, whereas Mo has been prepared by e-beam evaporation [5] or sputtering [20]. At the Air Force Research Laboratory, we observed the structure properties of MoS_2 films grown by sulfurization of DC magnetron sputtered MoO_3 and Mo precursor films at room temperature. In addition, reduced graphene oxide (rGO), known for increasing layer and domain size of MoS_2 [21, 22], was incorporated in our growth process of MoS_2. This chapter will be focused on our reported data related to this work.

S. Pacley (✉)
Materials and Manufacturing Directorate, Air Force Research Laboratory,
Wright-Patterson Air Force Base, OH, USA
e-mail: shanee.pacley@us.af.mil

© Springer Nature Switzerland AG 2020   49
M. E. Kinsella (ed.), *Women in Aerospace Materials*, Women in Engineering
and Science, https://doi.org/10.1007/978-3-030-40779-7_4

4.2 MoS$_2$ Research

Thin films of metallic Mo and MoO$_3$ were sputtered on c-face sapphire substrates (diameter of 25.4 mm) using a DC magnetron sputtering system (500 V at 100 mA) at room temperature, with an argon pressure of 0.92 Pa. The thickness of the precursor (3 nm for both Mo and MoO$_3$) was controlled by manipulating the sputtering so that there were equal amounts of Mo sputtered in the MoO$_3$ and Mo films. Table 4.1 lists the precursors and sample names. The substrates were ultrasonically cleaned in acetone for 5 minutes prior to deposition of Mo and MoO$_3$. Following sputtering of Mo and MoO$_3$ onto the substrates, the precursors were separately placed in the center of the quartz tube (Fig. 4.1). Sulfur powder (2 g) was placed in a ceramic boat, upstream from the Mo and MoO$_3$ films. Reduced graphene oxide (Sigma-Aldrich) was dispersed in isopropyl alcohol and drop cast on separate sapphire substrates. The rGO samples were air dried before they were placed in the furnace next to the sputtered precursor films of Mo and MoO$_3$ (with a distance of 5 mm between the precursor and rGO samples). After pumping the furnace down to a vacuum pressure of 667 Pa, the samples were heated to 300 °C at 20 °C/min and held there for 15 minutes. This enabled the removal of any residual water molecules. Next, the

Table 4.1 List of thicknesses used for Mo and MoO$_3$ precursor films

Sample (r-rGO)	Precursor	Precursor thickness
S1, S1r	Mo	3 nm
S2, S2r	MoO$_3$	3 nm

Reproduced from Pacley et al. [3], with the permission of the American Vacuum Society
r indicates rGO was used during experiments

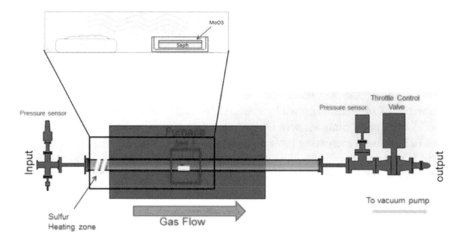

Fig. 4.1 Chemical vapor deposition setup for MoS$_2$ growth on Al$_2$O$_3$ substrates. (Reproduced from Pacley et al. [3], with the permission of the American Vacuum Society)

precursors were heated to 850 °C at a rate of 20 °C/min. As the temperature of the furnace approached 850 °C (around 830 °C), the boat with sulfur was heated to 125 °C using a heating tape. Both the precursors and the sulfur were held at their temperatures for 1 hour, followed by cooling to room temperature. All experiments were performed in an Ar/H_2 environment, with a flow rate of 75 sccm.

Transmission electron microscopy (TEM) imaging of the MoS_2 film cross sections for samples S1 (MoS_2 grown from Mo precursor) and S1r (MoS_2 film grown using Mo precursor with rGO seed) are shown in Fig. 4.2 [3]. The precursor films, MoO_3 and Mo, are both shown in Fig. 4.2a, b [3]. Samples S1 and S1r (Fig. 4.2c, e) show uniform and continuous layer growth of MoS_2 [3]. Both samples have a thickness of 7–8 nm, indicating the rGO used during the CVD growth of sample S1r had no effect on the film thickness. Atomic force microscopy (AFM) showed that samples S1 and S1r had an RMS of 0.360 nm and 2.43 nm (respectively), and the grain size increased from 4.5 nm to 17.7 nm, respectively (see Fig. 4.3a, b) [3]. This increase in the grain size indicated that the rGO played a role in grain growth of the MoS_2. In contrast to the uniform and continuous film growth of samples S1 and S1r, samples S2 (MoO_3 precursor) and S2r (MoO_3 precursor with rGO seed) demonstrated a non-uniform, outward growth of MoS_2 (Fig. 4.2d, f) [3]. It is reported that at 600 °C, MoO_3 reduces to MoO_2 under an H_2 environment [23]. In this research, there was indication that MoO_2 had formed after annealing MoO_3 at 850 °C. X-ray photoelectron spectra (Fig. 4.4a) showed peaks at 229.57 and 232.7 for Mo(IV), which is typical of MoS_2 and MoO_2, and 232.19 and 235.32 for Mo(VI), which is typical of MoO_3. AFM was performed on the same annealed sample (Fig. 4.3c) [3], and we noticed small islands across the substrate. The islands were formed when the sputtered MoO_3 film reduced to MoO_2 during annealing at 850 °C. Consequently, sulfurization of MoO_2 islands caused MoS_2 growth in a Volmer-Weber growth mechanism, which is a result of the film not wetting the substrate [24]. Moser and Levy reported similar growth patterns using sputtering technique to deposit thick MoS_2 films [25].

Figure 4.3d, e shows the grain structures of MoS_2 grown using the sputtered MoO_3 films (S2 and S2r) [3]. The RMS values for these samples were 2.00 nm (S2) and 3.66 nm (S2r), and the grain size increased from 7.9 nm (S2) to 12.2 nm (S2r) when rGO was used during the growth. This correlates well with the data from samples S1 and S1r that suggest rGO promotes grain growth when using sputtered precursor films. There was also a decrease in the film thickness, going from 15 nm (S2) to 7 nm (S2r) when rGO was used during the sulfurization process (Fig. 4.2d, f). Ling et al. [22] report that organic seed promoters (such as PTAS) enable heterogeneous nucleation sites and that the size of the MoS_2 domains is dependent upon the distance of the seed promoter from the precursor. We believe this is what occurred when rGO was used in our experiments involving sputtered Mo and MoO_3 films. However, further investigation needs to be conducted to better understand the kinetics, and mechanism of increasing grain size, when using rGO during the sulfurization sputtered films.

X-ray photoelectron spectroscopy (XPS) was performed for composition and chemistry analysis of the films that were grown in this research. The survey spectra

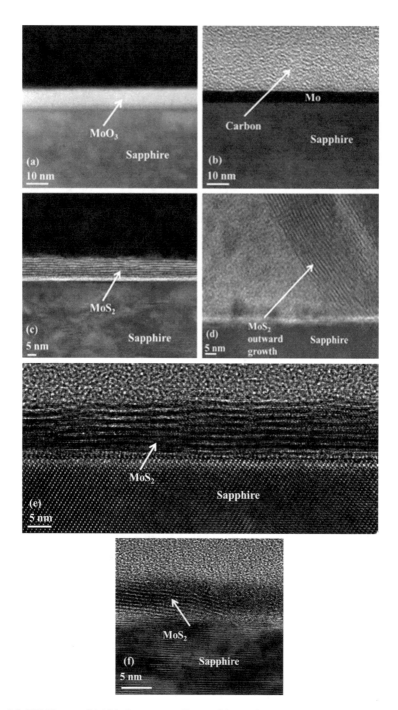

Fig. 4.2 TEM image of (**a**) MoO_3 precursor film used for MoS_2 growth; (**b**) Mo precursor film for MoS_2 growth; (**c**) sample S1 (MoS_2 on sapphire using Mo precursor) showing a layer thickness of 7 nm; (**d**) sample S2 (MoS_2 on sapphire using a MoO_3 precursor) showing an outward growth of MoS_2, with a thickness of 15 nm; (**e**) sample S1r (Mo precursor) using rGO with a measured thickness of 7–8 nm; and (**f**) sample S2r (MoO_3) using rGO with a thickness of 7 nm. (Reproduced from Pacley et al. [3], with the permission of the American Vacuum Society)

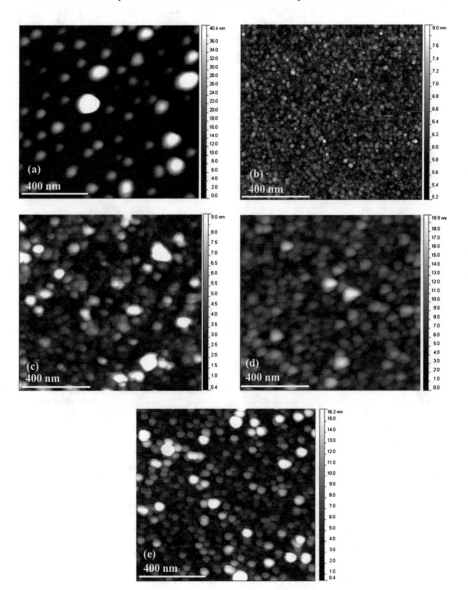

Fig. 4.3 AFM surface topography (1 × 1 μm) for (**a**) MoO3 precursor that was heated to 850 °C forming MoO$_2$ islands; (**b**) sample S1 (Mo precursor) showing a dense film of MoS$_2$ with a grain size of 4.4 nm; (**c**) MoS$_2$ sample S2 (MoO$_3$ precursor) with a grain size of 7.9 nm; (**d**) MoS$_2$ of MoS$_2$, with a thickness of 15 nm; (**e**) sample S1*r* (Mo precursor) using rGO with a measured thickness of 7–8 nm; and (**f**) sample S2r (MoO$_3$) using rGO with a thickness of 7 nm. (Reproduced from Pacley et al. [3], with the permission of the American Vacuum Society)

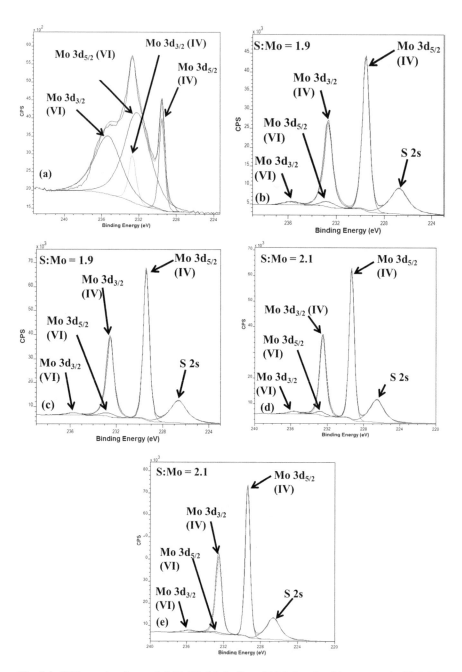

Fig. 4.4 XPS spectra of annealed MoO3, MoS$_2$ films S1(MoS$_2$ using Mo precursor), S2 (MoS$_2$ using MoO$_3$ precursor), S1*r* (MoS$_2$ using Mo precursor and rGO), and S2*r* (MoS$_2$ using MoO$_3$ precursor and rGO). The annealed MoO$_3$ (**a**) shows Mo(IV) peaks which are indicative of MoO$_2$ and Mo(VI) peaks that occur when MoO$_3$ is present. Both (**b**) and (**c**) show spectra for samples S1 and S2, respectively, having a stoichiometric structure. In (**d**) and (**e**), the stoichiometry has increased for samples S1*r* and S2*r*, respectively, indicating the rGO played a role in increasing the stoichiometry. (Reproduced from Pacley et al. [3], with the permission of the American Vacuum Society)

(not shown) from sulfurized thin films of samples S1/S1r and S2/S2r showed peaks from Mo and S, as expected. As mentioned previously, Fig. 4.4a shows the spectra for annealed MoO_3 [3]. The influence of the rGO on MoS_2 stoichiometry was analyzed by comparing the S:Mo ratios obtained from the XPS spectra. The MoS_2 films grown from samples S1 and S1r had S:Mo ratios of 1.9 and 2.1, respectively (see Fig. 4.4b, d). Samples S2 and S2r demonstrated the same respective S:Mo ratios of 1.9 and 2.1 (see Fig. 4.4c, e). The apparent improvement in the film stoichiometry is most likely a result of the Mo:S averaging over large spot size analysis area in XPS, which is orders of magnitude larger when compared to the average grain sizes of synthesized MoS_2 films. The presence of the rGO helped to increase the MoS_2 grain size areas and correspondingly decrease contributions of photoelectrons escaped from the grain boundary areas.

Curve fits to the Mo 3d doublets for all of the samples revealed two populations of Mo atoms. The Mo $3d_{5/2}$ peak at 229.8 eV and Mo $3d_{3/2}$ peak at 232.9 eV reveal the presence of Mo(IV), with a binding energy typical of MoS_2 or MoO_2 [26]. The Mo $3d_{5/2}$ peak at 232.7 eV and Mo $3d_{3/2}$ peak at 235.8 eV indicate the presence of Mo(VI), with a binding energy typical of MoO_3 [27]. This suggests that there is likely some MoO_2 present at the surface or at grain boundaries within the films. However, the intensity for both the Mo(IV) and Mo(VI) peaks are so low, that the presence of MoO_2 and MoO_3 is negligible.

4.3 Conclusion

The influence of metallic Mo and MoO_3 thin-film precursors on the structure of MoS_2 films grown by CVD was investigated. TEM established that rGO did not have an impact on the MoS_2 film thickness for sputtered Mo but that it was responsible for the increase in the grain size. We also observed an increase in the grain size when rGO was used during sulfurization of sputtered MoO_3. Reports demonstrate that seed promoters diffuse onto growth substrates, acting as nucleation sites for MoS_2 growth. In addition, the size of the MoS_2 domains is dependent upon the distance between the seed promoter and the growth substrate. In conclusion, sputtered Mo precursor films produce better uniformity and continuous MoS_2 films, making these nanocrystalline films potentially applicable for electronic and optoelectronic devices.

References

1. B. Radisavljevic, A. Radenovic, J. Brivio, et al., Single-layer MoS_2 transistors. Nat. Nanotechnol. **6**, 147 (2011)
2. Z. Yin, H. Li, H. Li, et al., Single-layer MoS_2 phototransistors. ACS Nano **6**(1), 74 (2012)

3. S. Pacley, J. Hu, M. Jespersen, et al., Impact of reduced graphene oxide on MoS_2 grown by sulfurization of sputtered MoO_3 and Mo precursor films. J. Vac. Sci. Technol. A **34**(4), 041505–041501 (2016)

4. D.Y. Zemlyanov, M. Jespersen, D.N. Zakharov, et al., Versatile technique for assessing thickness of 2D layered materials by XPS. Nanotechnology **29**(115705), 1 (2018)

5. Y. Zhan, Z. Liu, S. Najmaei, et al., Large-area vapor-phase growth and characterization of MoS_2 atomic layers on a SiO_2 substrate. Small **8**(7), 966 (2012)

6. S. Najmaei, Z. Liu, X. Zou, et al., Vapour phase growth and grain boundary structure of molybdenum disulphide atomic layers. Nat. Mater. **12**(8), 754 (2013)

7. B. Radisavljevic, M.B. Whitwick, A. Kis, Integrated circuits and logic operations based on single-layer MoS_2. ACS Nano **5**(12), 9934 (2011)

8. K.F. Mak, C. Lee, J. Hone, et al., Atomically thin MoS_2: A new direct-gap semiconductor. Phys. Rev. Lett. **105**, 136805–136801 (2010)

9. J. Brivio, D.T.L. Alexander, A. Kis, Ripples and layers in ultrathin MoS_2 membranes. Nano Lett. **11**(12), 5148 (2011)

10. H. Ramakrishna Matte, A. Gomathi, A. Manna, et al., MoS_2 and WS2 analogues of graphene. Angew. Chem. Int. Ed. **49**(24), 4059–4062 (2010). https://doi.org/10.1002/anie.201000009

11. Z. Zeng, Z. Yin, X. Huang, et al., Single-layer semiconducting nanosheets: High-yield preparation and device fabrication. Angew. Chem. Int. Ed. **50**(47), 11093–11097 (2011). https://doi.org/10.1002/anie.201106004

12. V. Nicolosi, M. Chhowalla, M.G. Kanatzidis, et al., Liquid exfoliation of layered materials. Science **340**(6139), 1226419 (2013). https://doi.org/10.1126/science.1226419

13. G. Eda, H. Yamaguchi, D. Voiry, et al., Photoluminescence from chemically exfoliated MoS_2. Nano Lett. **11**(12), 5111–5116 (2011). https://doi.org/10.1021/nl201874w

14. C. Muratore, A.A. Voevodin, Control of molybdenum disulfide basal plane orientation during coating growth in pulsed magnetron sputtering discharges. Thin Solid Films **517**(19), 5605–5610 (2009). https://doi.org/10.1016/j.tsf.2009.01.190

15. C. Muratore, J.J. Hu, B. Wang, et al., Continuous ultra-thin MoS_2 films grown by low-temperature physical vapor deposition. Appl. Phys. Lett. **104**, 261604 (2014). https://doi.org/10.1063/1.4885391

16. K. Kang, S. Xie, L. Huang, et al., High-mobility three-atom-thick semiconducting films with wafer-scale homogeneity. Nature **520**, 656 (2015)

17. C. Altavilla, M. Sarno, P. Ciambelli, A novel wet chemistry approach for the synthesis of hybrid 2D free-floating single or multilayer Nanosheets of MS_2@oleylamine (M=Mo, W). Chem. Mater. **23**(17), 3879–3885 (2011). https://doi.org/10.1021/cm200837g

18. H. Schmidt, S. Wang, L. Chu, et al., Transport properties of monolayer MoS_2 grown by chemical vapor deposition. Nano Lett. **14**(4), 1909–1913 (2014). https://doi.org/10.1021/nl4046922

19. Y. Yu, C. Li, Y. Liu, et al., Controlled scalable synthesis of uniform, high-quality monolayer and few-layer MoS_2 films. Sci. Rep. **3**, 1866 (2013)

20. N. Choudhary, J. Park, J.Y. Hwang, et al., Growth of large-scale and thickness-modulated MoS_2 nanosheets. ACS Appl. Mater. Interfaces **6**(23), 21215–21222 (2014). https://doi.org/10.1021/am506198b

21. Y. Lee, X. Zhang, W. Zhang, et al., Synthesis of large-area MoS_2 atomic layers with chemical vapor deposition. Adv. Mater. **24**(17), 2320–2325 (2012). https://doi.org/10.1002/adma.201104798

22. X. Ling, Y. Lee, Y. Lin, et al., Role of the seeding promoter in MoS_2 growth by chemical vapor deposition. Nano Lett. **14**(2), 464–472 (2014). https://doi.org/10.1021/nl4033704

23. E. Lalik, W.I.F. David, P. Barnes, et al., Mechanisms of reduction of MoO_3 to MoO_2 reconciled. J. Phys. Chem. B **105**(38), 9153–9156 (2001). https://doi.org/10.1021/jp011622p

24. D.L. Smith, *Thin Film Deposition: Principles and Practice*, 1st edn. (McGraw-Hill Education, New York, 1995)

25. J. Moser, F. Levy, Growth mechanisms and near-interface structure in relation to orientation of MoS_2 sputtered thin films. J. Mater. Res. **7**(3), 734–740 (1992). https://doi.org/10.1557/JMR.1992.0734
26. G. Seifert, J. Finster, H. Müller, SW Xα calculations and x-ray photoelectron spectra of molybdenum(II) chloride cluster compounds. Chem. Phys. Lett. **75**(2), 373–377 (1980). https://doi.org/10.1016/0009-2614(80)80534-3
27. P.A. Spevack, N.S. McIntyre, Thermal reduction of molybdenum trioxide. J. Phys. Chem. **96**(22), 9029–9035 (1992). https://doi.org/10.1021/j100201a062

Chapter 5
Emerging Materials to Move Plasmonics into the Infrared

Monica S. Allen

5.1 Introduction

The field of plasmonics combines many attractive features in nanoelectronics and optics and opens the door to highly integrated, dense subwavelength optical components and electronic circuits that will help alleviate the speed bottleneck in important technologies such as information processing and on-board computing. Plasmonic devices may also be able to provide improved performance with respect to cost, size, weight, and power consumption as compared to conventional systems. However, the wide application of plasmonic devices hinges on practical demonstrations with low losses at standard optical wavelengths, such as visible, telecom, and IR. Plasmonics has received much interest in recent years, and several research efforts have been aimed at understanding the underlying physics [1]. Recent work has focused on demonstrations that overcome the fabrication and material constraints and enable maturation of the technology into field-testable applications, such as optical computing and chips and enhanced signal detectors [2]. Surface plasmon polaritons (SPPs) are quasi-particles or excitations that result from resonant coupling of photons to the collective oscillations of conduction electrons in materials [3]. SPPs can be described as two-dimensional (2D) bound electromagnetic waves that propagate along conductor-dielectric boundaries [4] and exponentially decay away from this interface [5, 6]. These waves are characterized by strongly enhanced localized fields and high spatial confinement, sometimes at subwavelength scales. These attributes make SPPs attractive for the design of components since they are not constrained by the diffraction limit [7] and may reduce size, weight, and power consumption when compared to conventional systems [1, 5].

Conventional plasmonic devices and arrays utilize metallic thin films or periodic arrays built with fabrication processes, such as electron beam lithography (EBL), to

M. S. Allen (✉)
Munitions Directorate, Air Force Research Laboratory, Eglin AFB, FL, USA
e-mail: monica.allen.3@us.af.mil

© Springer Nature Switzerland AG 2020
M. E. Kinsella (ed.), *Women in Aerospace Materials*, Women in Engineering and Science, https://doi.org/10.1007/978-3-030-40779-7_5

provide gain by coupling the optical signal to the plasmonic modes. Metals suffer from high losses at optical and infrared (IR) wavelengths that are difficult to compensate for completely by simply adding gain material, making the implementation of practical plasmonic-based devices challenging at these frequencies. Some research groups have explored adding optical gain using additional active medium, such as pumped dyes in the dielectric layer, to compensate for internal, radiative, and total losses in the plasmonic waveguides [8–11]. Another field that has been based on plasmonic phenomena is transformation optics (TO). This methodology enables devices, such as hyper- and super-lenses, and negative or near-zero refractive index with unprecedented functionality. However, practical demonstrations of TO-based devices have been restricted to the ultraviolet (UV) regime where conventional plasmonic materials like Ag have lower losses than in the visible or IR [12–14]. Apart from large losses, metals also restrict practical demonstrations of TO devices at these frequencies due to large magnitudes of negative real permittivity [15]. The requirement of low permittivity values stems from the fact that many TO-based devices need the magnitude of the permittivity in the metal to be comparable in magnitude and opposite in sign to that of the dielectric component so that their responses are balanced. In the optical regime, dielectrics usually have permittivity values lower than 10. Thus, metals with large negative real permittivity values at optical frequencies are not suitable for these metamaterial-based devices. Both issues of permittivity values and losses can be overcome using alternate materials that will enable the transition of these devices to the IR.

5.2   Material Properties

The values of mobility and carrier concentration can be used in the Drude model of dielectric function to predict the optical properties associated with plasmonic films. The Drude-Lorentz function can be defined such that the permittivity is expressed as a function of frequency as described by Eq. 5.1. The Drude model accounts for free carriers and the Lorentz oscillator for the interband transitions at the band edge. The square root of the carrier concentration directly relates to the plasma frequency, and thus the plasmon resonance frequency can be modified by changing the carrier concentration [16, 17].

$$\varepsilon(\omega) = \varepsilon_{\mathrm{r}} + i\varepsilon_{\mathrm{i}} = \varepsilon_{\mathrm{int}} - \frac{\omega_{\mathrm{p}}^2}{\left(\omega^2 + \gamma^2\right)} + i\frac{\gamma\omega_{\mathrm{p}}^2}{\omega\left(\omega^2 + \gamma^2\right)}$$

$$= \varepsilon_{\mathrm{int}} - \frac{\left(n_{\mathrm{opt}}e^2 \middle/ \varepsilon_0 m^*\right)}{\left(\omega^2 + \gamma^2\right)} + i\frac{\gamma\left(n_{\mathrm{opt}}e^2 \middle/ \varepsilon_0 m^*\right)}{\omega\left(\omega^2 + \gamma^2\right)} \qquad (5.1)$$

where ε_r and ε_i are the real and imaginary part of the permittivity, respectively, ε_{int} is the intrinsic permittivity of the undoped material, n_{opt} is the carrier concentration, ω is the operating frequency, e is the electron charge, $m*$ is the effective electron mass, and γ is the Drude relaxation rate which can be approximated by $\gamma = e/\mu_{opt}m*$ where μ_{opt} is the optical mobility. The real part of the permittivity (ε_r) determines interaction with incident or propagating waves and affects the polarization induced by incident or applied electric field. The imaginary part of the permittivity (ε_i) determines losses experienced in the media and is also known as the extinction coefficient. Materials with negative ε_r are required for plasmonics, and thus appropriate materials need a plasma frequency greater than the operating frequency. Thus, the ideal plasmonic material would be lossless ($\varepsilon_i = 0$) and have a real and negative permittivity ($\varepsilon_r < 0$) which is not practically achievable.

Metals are generally limited to the UV and visible spectral ranges since plasma frequency, ω_p, is proportional to $n^{1/2}$, where n is carrier concentration, and it is difficult to change carrier concentration in a metal with moderate external stimulus, such as electric or optical field or temperature. Thus, devices based on metals are not easily tunable and have to be designed to work at a specific frequency; and multiple designs are needed for different operating wavelengths, which is not ideal for multispectral applications [2]. The next consideration is that of growth and fabrication limitations. While the most prevalent material for plasmonics is gold, because of its low losses and relative inertness, silver, aluminum, and copper have also been used, although they quickly react in air to form compounds. Alkali metals such as sodium and potassium have very low losses but cannot be used practically since they are highly reactive to air and water, and maintaining the elemental form that offers these attractive properties is very challenging. Even in the case of noble metals, the very thin films required for plasmonics are difficult to grow using standard techniques, such as evaporation and sputtering. These techniques give discontinuous films with high surface roughness compared to the overall film thickness for very thin films, which results in increased scattering and degraded optical properties. These non-uniformities are exacerbated by processing, such as patterning, and can increase the relaxation rate by as much as three- to five-fold and, in turn, increase losses. Therefore the search for suitable candidates for plasmonic materials is dependent not only on their predicted material properties but also on practical considerations such as ease of handling and processing. Also, as the technology matures, manufacturing considerations and scalability should be a factor in performance metrics and viability determinations [18].

Many emerging materials hold the promise of tailoring material properties, in addition to device geometry, to tune for different regions of the spectrum. They also offer several advantages including the possibility of bandgap engineering, tunability, selectivity, and greater compatibility with semiconductor deposition processes, such as chemical and physical vapor deposition (e.g., pulsed laser deposition, sputtering, etc.). Conducting materials can be grown to have appropriate carrier concentrations and thus smaller losses. Some of these materials have demonstrated significantly lower losses when compared to silver, the metal that exhibits the lowest losses in the visible (VIS) and near-IR ranges. This chapter reviews the wide range of available plasmonic materials for IR wavelengths that make good candidates to replace metals, including transparent conducting oxides, such as doped zinc oxide and indium tin oxide; transition metal nitrides, such as titanium nitride; doped semiconductors, such as gallium phosphide and nitrides; organic conductors, such as graphene and conductive polymers; ceramics and newer material platforms, such as perovskites; and unique alloys and combined nanostructures, such as core-shell nanoparticles.

5.3   Transparent Conducting Oxides

There are several oxides that have been explored for plasmonic materials. Some examples are tin oxide (SnO_2), indium tin oxide (ITO), zinc oxide (ZnO), vanadium dioxide (VO_2), indium oxide (In_2O_3), and some copper- and cadmium-based compounds. These oxides can be degeneratively doped to achieve carrier concentrations of $\sim n \geq 10^{21}$ cm^{-3}, which translates to $\lambda_p \sim 1$ µm, opening up the IR region with much lower losses than metals at the same wavelength [19]. In addition to doping oxides, various combinations of these oxides have been explored such as $\{ZnO\}_{0.05}$:$\{SnO_2\}_{0.05}$:$\{In_2O_3\}_{0.9}$ (ZITO)   and $\{In_2O_3\}_{0.05}$:$\{SnO_2\}_{0.05}$:$\{ZnO\}_{0.9}$ (ITZO) [16]. These oxides and other compounds have large bandgaps, are therefore transparent in the visible wavelength regime, and are thus collectively often referred to as transparent conducting oxides (TCOs). TCOs can be grown into crystalline and polycrystalline structures. TCO thin films can be easily patterned using standard lithographic approaches to obtain subwavelength structures that are often needed in plasmonic devices. TCOs can be grown by many different methods, such as pulsed laser deposition (PLD) and sputtering, and their resonances can be tuned by controlling carrier concentration (n) and, to a lesser extent, mobility (μ). As a practical example, PLD growth of ZnO with simple air anneals has been reported to hit the telecommunication wavelengths of 1.3 and 1.55 µm. Some research has even explored growing these films with sol-gel techniques with lesser success on repeatability and film quality [20–22]. On the other hand, these sol-gel techniques are inexpensive from an infrastructure standpoint and are easily scaled for larger areas. Further research may improve the yields and quality of films that have been demonstrated thus far.

TCOs have been extensively researched for electrodes and find applications in liquid crystal displays, solar cells, and light-emitting diodes [23]. The most commonly utilized TCO is indium tin oxide (ITO); however, it has the following disadvantages. The cost of ITO thin films is prohibitive because indium is a scarce element and ITO thin films are vapor-deposited at very slow rates. Also, ITO is brittle and susceptible to mechanical damage, rendering it unsuitable for applications such as flexible solar panels [24, 25]. A strong candidate to replace ITO is ZnO doped with Al, Ga, or In. Al-doped ZnO (AZO) has also recently been proposed as a plasmonic material in the IR region, a range not accessible to metal-based plasmonics [26–29]. AZO can be easily fabricated into waveguides using standard fabrication techniques, e.g., photolithography, and patterned into periodic structures such as conductor-hole arrays and nanowire arrays [30–33]. Ga-doped ZnO (GZO) using PLD with simple air anneals can be used to red shift the plasmonic resonance to IR wavelengths of 1 μm and beyond [34–36]. However, one potential problem with TCO thin films is that most of them are grown on lattice-mismatched substrates, producing poor material near the interface. In these cases, the mobility in this region is reduced, and thus the overall electrical and optical properties are affected and dependent on layer thickness [37–39]. Since uniform thin films with low scattering losses are key to producing practical plasmonic devices, it is not only necessary to understand the origins of this thickness dependence but also to overcome it using careful design of growth processes [40–42]. Itagaki et al. demonstrated improved crystallinity of RF-sputtered ZnO by inserting a thin nitrogen-mediated zinc oxide (ZnON) buffer layer between the substrate and ZnO layer. For example, in undoped ZnO grown on c-plane Al_2O_3, the rocking-curve full width half maximum (FWHM) of the (002) diffraction dropped from 0.490° to 0.061°, and in AZO grown on quartz glass substrates, the thickness dependencies of ρ, μ, and n were greatly minimized. These uniform films also displayed low surface roughness and reduced defect concentrations, leading to lower losses [43]. The buffer layers shown in the paper are 20 nm thick, but layers as thin as 5 nm give similar results. Electrically, the buffer layers are semi-insulating with minimal effect on the conductivity of the AZO film. The purpose of the nitrogen in the buffer layer is to inhibit the strong nucleation tendency of ZnO, which leads to small grain sizes and results in larger scattering. It has been shown that the use of a buffer layer can increase grain sizes from 38 to 68 nm, measured using X-ray diffraction ω-2θ scans. The buffer also reduces the thickness dependence of μ and n [41–43].

The electrical properties of TCOs can be derived from four-point probe measurements, and the optical properties can be measured using ellipsometry and reflectance measurements [40, 44, 45]. These parameters can then be used in the Drude-Lorentz model, described in the previous section, to derive properties of interest [46–48]. Other complex oxides, including perovskites, have also been suggested as alternative plasmonic materials. Some examples are strontium titanate, barium titanate, barium strontium titanate, strontium stannate, barium stannate, strontium stannate, cadmium tellurate, etc. These complex oxides offer additional functionality such as switching and modulation [17, 49]. Further, phase transition materials that switch from insulator to metal phase such that the solid material goes from electrically nonconductive to conductive have recently been explored as candidate materials for plasmonics. These materials, e.g., vanadium dioxide, have large changes in refractive index when switched between covalent and resonant bonding states. The materials can open the door to additional functionality in niche applications [50].

5.4 Metal Nitrides

Another class of materials that has been explored as alternatives to metals in plasmonic applications is metal nitrides and transition metal nitrides (TMNs), which form a special category of ceramics. TMN materials have very high melting points and, similar to TCOs, have high carrier concentrations and mobility values. Additionally, their material properties can be tuned by changing chemical composition. These TMNs have been researched from near UV to VIS to IR [51–54]. Figure 5.1 shows that TMNs can have tunable plasmonic activity in the IR range [51]. Ternary metal nitrides such as titanium, zirconium, hafnium, and tantalum-based nitrides can be tuned from the UV to visible region of the spectrum. These nitrides can be alloyed with elements from group III (Sc, Y, Al) or group II (Mg, Ca) with two or three valence electrons to red shift the plasmon resonance to the IR regime by forming ternary or, more accurately, pseudobinary nitrides (B1 structure is maintained). An additional advantage is the CMOS compatibility of these nitrides. They can be grown into high-quality films on a variety of substrates such as silicon, c-sapphire, and magnesium oxide using techniques such as sputtering, chemical vapor deposition (CVD), atomic layer deposition (ALD), PLD, and ion beam-assisted deposition (IBAD) [55, 56]. Not only have transition metal nitride films been explored, but recent work has also examined the use of TMN nanoparticles for plasmonics. These provide higher efficiency than noble metals, such as gold, in the near infrared [51], which is a region that is not only important in defense but also in biological imaging, due to optical transparency of biological tissue. This biological

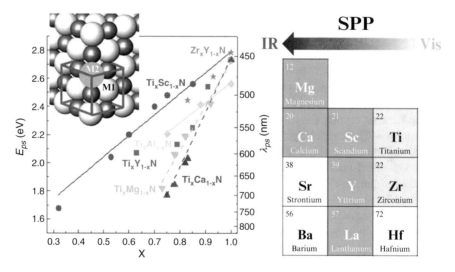

Fig. 5.1 Conductive transition metal nitrides are emerging as promising alternative plasmonic materials that are refractory and CMOS-compatible. This work shows that ternary transition metal nitrides of the B1 structure, consisting of a combination of group IVb transition metals, such as Ti or Zr, and group III (Sc, Y, Al) or group II (Mg, Ca) elements, can have tunable plasmonic activity in the IR range [51]

transmission window reduces losses and allows greater depth for imaging. Additionally, nitride nanoparticles provide the opportunity for local heating using the absorption processes in the particles [55].

Recently, doped gallium, indium, and aluminum nitride have been explored as possible candidates for plasmonics in the far IR and terahertz region. Nitrides such as GaN are attractive due to their chemical resistance and non-toxic nature, which makes them ideal for sensing. For example, InN has an inherent surface electron accumulation with low electron mass and high saturation velocity, which gives it superior electronic properties [57]. However, there are fabrication challenges with these materials due to their low melting points. Both epitaxially grown thin films and nanostructures have been researched. Melentev et al. have shown that wurtzite GaN films deposited with metal-organic vapor phase epitaxy (MOVPE) can be heavily doped with silicon donor impurities to obtain high carrier concentrations of $\sim 3.6 \times 10^{19}$ cm^{-3} and mobilities of ~ 120 cm^2/Vs at room temperature [58, 59]. Random nanostructures of InN and In nanoparticles in InN matrix have both been investigated for plasmonic properties. The wavelengths of the SPPs for intrinsic InN are in the range of 275–500 nm (UV), which can be attributed to the 2D electron gas. However, the doping concentration can be changed to vary the carrier density and thus the plasmonic properties [57]. Further, wurtzite GaN epitaxial layers heavily doped with silicon donor impurities on planar and patterned surfaces have also been examined and reported to provide a constant free electron concentration with high mobility, resulting in extremely narrow and deep resonances in the terahertz range. There also exist low-frequency plasmon-phonon modes that overlap energetic carriers and elastic waves, which take their functionality one step further.

5.5 Semiconductors

Semiconductors present an important class of materials for extending plasmonic frequencies into the mid- and far-IR when doped at appropriate levels to tune the resonance. Most semiconductors can be grown epitaxially using molecular beam epitaxy (MBE) or metal-organic chemical vapor deposition (MOCVD) and doped using ion implantation or diffusion. These growth techniques produce high-quality films with single crystalline materials and low defects. Also, these materials are already used in CMOS chips, which further enhances their utility due to well-studied compatibilities in lattice structures and methods for mitigating mismatches. Additionally, the processes for growth and fabrication are very refined and established, thus allowing precise control over doping (free carrier concentration) and mobility, both key parameters for determining surface plasmon resonances [60]. In some studies, intense optical excitation with photon energy higher than the semiconductor bandgap has been used to increase free carrier concentration and thus modify the plasma frequency in real time. However, ultrahigh doping has challenges in all the materials discussed thus far, namely, semiconductors, oxides, and nitrides. These include:

(i) *Doping efficiency*: Interstitial doping does not affect free carrier concentration, only substitutional doping does. Therefore, the type of doping matters.

(ii) *Solid solubility limit*: This is the thermodynamic upper limit on the amount of dopant that can be absorbed by a material and can be thought of as a doping saturation point.

(iii) *Secondary phases*: Very high doping concentrations can result in secondary unwanted phases.

(iv) *Crystalline defects*: Very high doping also creates crystal defects that present traps for free carriers and adversely affect performance.

(v) *Effect of mobility*: High doping concentration causes the relaxation rate to increase which in turn reduces mobility. Thus, there is a definite trade space to be considered.

Semiconductor materials, such as silicon and germanium (group IV), cadmium/zinc (group II) alloyed to form selenide and tellurides (group VI), and gallium/indium (group III) alloyed to form phosphides and arsenides (group V), have been explored for moving plasmonic resonances in the infrared frequency range [61, 62]. GaN also can fall into this category but is discussed with nitrides in the previous section. Some demonstrations of these materials include doped GaAs and InAs that have high carrier concentrations and exhibit surface plasmonics for wavelengths up to 9 μm [63, 64]. Combinations of III–V compounds in material systems, such as GaAs and AlGaAs, have been demonstrated at telecommunication wavelengths of 1.3 μm and 1.55 μm. Germanium is a material which provides a plasmon resonance in the mid-IR when heavily doped. This is due to the lack of optical phonon absorption, which leads to low losses in the mid-IR, and the fact that the plasma frequency can be modulated using optically generated electron-hole pairs. It is also attractive for nonlinear plasmonics because of its high third-order nonlinear coefficient [65]. Doped silicon (both n and p) has also been used effectively in the IR regime. Recently, research has also been done in antimonide-based plasmonic devices, but these materials are primarily used in the long-wave IR and terahertz frequencies. Furthermore, semiconductors have been patterned and fabricated into devices and shown to have good performance as plasmonic detectors, modulators, waveguides, etc. [66].

5.6 Graphene and Organics

Two promising candidates for non-metallic carbon-based materials for plasmonics are graphene and conducting polymers. Graphene offers unique features when compared to most of the materials discussed thus far; for example, it has intrinsic plasmonic resonances that are tunable with low dissipation [67, 68]. This is primarily because it is a 2D material where the film is a monolayer of tightly packed carbon atoms arranged in a honeycomb structure. Graphene has high conductivity coupled with high mobility and can be doped to achieve plasma resonance frequencies in the mid-IR wavelength regime. Even though it is a 2D material, graphene has been

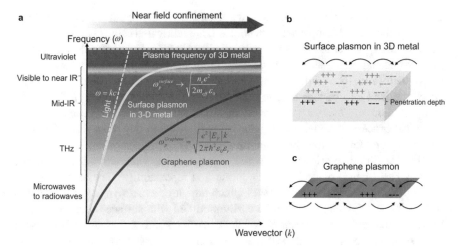

Fig. 5.2 (**a**) Sketch of dispersion curves for light wave, surface plasmonic wave at the interface between 3D noble metal and air, and graphene plasmonic wave. In the figure, m_{eff} is the effective mass of electron in 3D metal, n_e is the volume density of electrons in 3D metal, ε_0 is the vacuum permittivity, ε_r is the average dielectric constant of the surrounding medium, $|E_F|$ is the Fermi level of graphene, $\hbar$ is the reduced Planck constant, and c is the speed of light. Illustration of the charge, electric field (arrows) associated with (**b**) surface plasmonic wave on 3D bulk metal and (**c**) with 2D plasmonic wave in graphene [67]

shown to have high spatial confinement of fields and large plasmon lifetime. It can be deposited using CVD and also combined with a variety of materials and patterned into structures, such as stacks, to enhance functionality. The comparison between a graphene-based plasmon and that in a metal is shown in Fig. 5.2, which shows the dispersion curves for both and illustrates the propagation of a surface plasmon polariton in a 2D graphene sheet compared to a metal thin film [67]. Graphene could well be the new building block for plasmonic devices, especially in the 2D. Conducting polymers or polymer composites have been used in various optical applications such as electrodes and photovoltaics. Research has shown that purely organic materials can support a surface plasmon mode at the material-air interface and demonstrate near-zero or negative permittivity [69–71]. Recently, organic materials, such as polymers doped with dye molecules, have been investigated for surface exciton plasmons [72], and strong laser pulse excitation has also been employed to elicit a nonlinear response that may find use in optical switching [73].

5.7 Nanoparticles and Nanostructures

This chapter has presented plasmonic materials thus far in the context of films. There is a lot of interesting research that has also looked at nanostructures and nanoparticles as a means of exciting and supporting plasmonic modes. In this chapter, nanostructures are primarily used to describe patterns that are defined using top-down methods, such as EBL, that are often used to define subwavelength structures on substrates. Nanoparticles, in contrast, are fabricated using chemical synthesis, and their properties are manipulated either within the synthesis process or post-processing steps, such as annealing. Both nanostructures and nanoparticles derive their properties from geometry and material composition. Nanoparticles can be synthesized into various shapes with different materials for plasmonics and are being studied for their versatility of material composition and geometry [74]. Chemical processes are used to synthesize particles, and this often results in non-uniformity and lower yield and reliability.

The question then arises, why explore nanoparticles when a top-down process, such as EBL, can provide better resolution and repeatability? The answer lies in the limitations of top-down processes which are slow, expensive, and not scalable to large areas. Nanoparticles can be easily modified with surface functionalization to agents that aid in self-assembly of these particles into arrays using bottom-up processes. Self-organization (self-assembly) can be viewed as achieving a state of equilibrium either due to thermodynamic factors or, in the case of externally directed self-assembly (e.g., using poling fields), as a charge equilibration state [75]. In both the cases, the basic principle that governs the behavior can be characterized by a local or global minimum in the free energy of the system and thus can be described by formulations similar to those applied in dynamic flow systems. Tailored assemblies, such as dimers, trimers, 1D chains, or 2D surfaces, can leverage the interparticle coupling and the properties that result from it. Also, these ensembles are based on surface chemistry of the particles themselves, are therefore not limited by area or size of array, and do not require specialized equipment or expensive infrastructure to execute large sample sizes required for practical applications [76, 77]. Particles can also be assembled via templating, in which a polymer can align particles, or freezing them once they are aligned by external poling fields. Poling of particles is especially important for polarization sensitivity. Additionally, nanoparticles can be dispersed into polymer matrices to create hybrid materials that cumulatively improve the properties of the system, such as mechanical stability or responsiveness.

Nanoparticles can be made from many materials [78, 79] and can even be built from a combination of materials. For nanoparticles made of single materials, gold has been extensively studied for plasmonics due to its chemical and physical stability and ease of surface functionalization [80]. The natural resonance of a gold sphere of diameter ~20 nm is in the range of 500–550 nm. This resonance can, however, be redshifted by changing the shape of the particle to make it more anisotropic. When nanorods are synthesized, they possess a transverse resonance of ~520 nm along the shorter radial axis and a longitudinal resonance corresponding to the electron oscillation along the longer axis that is determined by the aspect length (length/width ratio) of the rod. Thus the wavelength of the longitudinal resonance can be controlled by the dimensions of the rod. Gold nanorods have been shown with resonances of ~1 μm, which required lengths of ~400 nm. These dimensions are difficult to synthesize and control and often have large non-uniformity and low yield [81]. Other single element colloidal nanoparticles such as magnesium [82] have also been

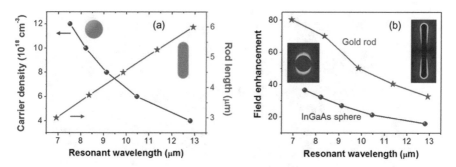

Fig. 5.3 (**a**) Tuning of SPP resonant wavelength obtained by adjusting doping in InGaAs sphere or the length of the Au rod with fixed cross section of 0.6 μm diameter. (**b**) Maximum achievable field enhancement for both structures [78]

explored for the NIR range with success. Alternative materials to noble metals similar to those discussed above such as conducting oxides, nitrides, and semiconductors have also been successfully synthesized into nanoparticles [83]. Figure 5.3 shows the comparison of the resonant wavelength tuning and field enhancement achieved at infrared wavelengths by adjusting the geometry of an Au rod vs. modifying the doping of an InGaAs nanoparticle [78]. As an example, nanoparticles and nanorods of ITO have both been shown to have plasma frequencies in the mid-IR [84, 85]. Similar to the gold nanorods, ITO nanorods have been grown on lattice-matched substrates. These arrays demonstrate the transverse and longitudinal plasmon resonances which can be tuned using post-processing steps, such as annealing [86]. The change in carrier concentration by annealing in a nitrogen-rich environment creates more oxygen vacancy defects and also helps with surface passivation, thus helping to redshift the plasma resonance frequency. Additionally, aluminum-doped zinc oxide particles have been demonstrated for wavelengths of ~2 μm using chemical doping during the synthesis process. Research has also demonstrated II–VI nanocrystals of telluride and selenides [87, 88], where the carrier density can be controlled via doping like the thin films. Among semiconductors, silicon nanocrystals can be doped for tunable plasmon resonances. More recently, Nordlander et al. have shown that silicon nanocrystals can be doped together with phosphorus and boron, allowing for highly tunable plasmon resonances. These resonances result from the contributions of phosphorus and boron to the light (transverse) and heavy (longitudinal) electrons and holes, respectively, giving rise to two plasmon branches. The interaction between these two branches results in an antibonding mode that is bright (higher energy, components in phase) and a bonding mode that is dark (lower energy, components out of phase) that can only be observed in the quantum regime [88]. Figure 5.4 shows examples of nanoparticles with varied shapes and materials and associated resonances.

More recently, biphasic core-shell particles have been investigated for plasmonic applications. These particles combine dissimilar materials with distinct properties in close proximity to enable unprecedented properties that cannot be obtained by either isolated material [90, 91]. Core-shell nanoparticles can be fabricated epitaxially for

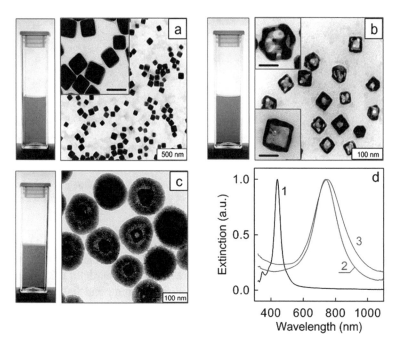

Fig. 5.4 TEM images and photos of samples with silver nanocubes (**a**), Au-Ag nanocages (**b**), and composite silica-coated nanocages (**c**). The insets in panel b illustrate the box and cage particle morphologies. Panel d shows the extinction spectra for Ag cubes (1), Au-Ag nanocages (2), and composite Au-Ag/SiO$_2$ particles (3). The scale bars in the insets are 50 nm [89]

materials with low to moderate lattice mismatch or with synthesis based on polymers where this limitation is overcome. The design space is also dramatically improved in core-shell nanoparticles since one can tune the properties by varying geometry (size, placement, aspect ratio, etc.), core/shell materials, or combinations of both [92, 93]. Both single-core, single-shell particles and single-core, multiple-shell particles with different shapes have been demonstrated. Asymmetric geometries where the centers of the core and shell are aligned differently to form nanoeggs (centers offset) and nanocups (core is offset and goes through the shell) have also been demonstrated with possible applications as tunable nanoscale lenses or tunable plasmonic nanostructures [94]. Halas et al. have shown that optical and magnetic properties can be combined into a nanoparticle with an iron oxide core that is surrounded with a gold shell [92]. Such nanoparticles could be assembled with an applied magnetic field into metasurfaces or superlattices to leverage the interaction between particles or build photonic-magnetic devices. Ziolkowski et al. designed nanoparticles with high index dielectric cores and semiconductor shells to simultaneously excite both electric and magnetic resonances with minimal backscattering or enhanced forward scattering through destructive and constructive interference effects in the far field [95]. Many exciting new avenues continue to be explored to push the limits of synthesis, assembly, and modification of nanoparticles. As these techniques mature, new avenues will be researched to transition these particles into viable applications that will require development of chemical methods to mass produce these nanoparticles. Filler et al. have researched this "nanomanufacturing" challenge and identified that an iterative sequence of synthesis and assembly, separation and sorting, and stabilization and packaging will be warranted to overcome the barriers to producing nanomaterials in large quantities [96].

5.8   Future Outlook

A key aspect of the alternative materials discussed in this chapter is that they are tunable, that is, their properties can be tailored to the application. For example, the plasmon resonance frequency can be tuned in wavelength or bandwidth and even formed into dual spectral peaks by tweaking the geometry and the composition of the materials. The electric field profiles and polarization sensitivity can also be controlled by patterning and assembly into predefined structures. The resonant nature of plasmonic materials and structures naturally lends itself to narrow band sensors required to discriminate spectral features, including threat warning, chemical identification, and multispectral sensing [97, 98]. Resonant nanoarrays can improve spectral selectivity and enable separation of many closely spaced spectral features. Additionally, the gain provided by coupling incident radiation to plasmonic excitation can significantly improve the signal to noise ratio (SNR) and speed of systems (sources or detectors), which is important in noisy environments. The high confinement of propagating SPPs enables subwavelength waveguides that can be used for on-chip optical interconnects for future all-photonic platforms [78, 99]. This will also help improve speed for information processing and on-board computing in systems with large data throughput [100].

From an economic standpoint, some of the materials discussed, e.g., zinc oxides, provide the advantage of significantly lower cost as compared to precious metals and fabrication methods that are compatible with standard semiconductor processing techniques, thus removing the need for significant investment in infrastructure and retooling. As the processes and theoretical models are developed and matured, there may be new capabilities for device enhancements (including real-time control), such as wavelength switching and tuning. As new materials are researched and fielded, especially nanoscale building blocks (like nanoparticles), it will be necessary to develop characterization techniques that can extract the wavelength dependent properties at the nanoscale. The properties at this scale are very different from the bulk material [101]. These methods will be even more important as research progresses to fabricate complex nanomaterials, such as single-core multiple-shell particles, with dielectric, metallic, and gain media all combined into a single particle at nanometer dimensions. The challenge is not simply to measure the optical and electrical properties but to isolate a single particle to measure its individual properties, because there will be coupling between adjacent layers that must be understood first before measuring the effect of organizing into arrays or ensembles. Analytical models are also being developed to define and predict the self-assembly of particles. Computational studies have been used to provide insight into the effect of various thermodynamic factors and external fields on the dispersion and alignment of nanorods in polymers [102]. Therefore, there are two distinct categories for application of plasmonics in the IR range: first, propagating plasmons, where the confinement and losses in waveguides for communication and information processing are examined, and second, devices and systems where the near-field coupling to surface plasmons is used for enhancement, shown recently in the enhancement of the performance of a broadband infrared FET with plasmonic overlays [103]. What really opens the realm of the possible in plasmonics is the use of the alternative materials discussed in this chapter that not only enable operation at longer wavelengths but also allow for careful tuning of material properties to best suit the application.

References

1. W.L. Barnes, A. Dereux, T.W. Ebbesen, Surface plasmon subwavelength optics. Nature **442**, 824–830 (2003)
2. S.A. Maier, M.L. Brongersma, P.G. Kik, S. Meltzer, A.A.G. Requicha, H.A. Atwater, Plasmonics—A route to nanoscale optical devices. Adv. Mater. **13**, 1501–1505 (2001)
3. R.H. Ritchie, Surface plasmons in solids. Surf. Sci. **34**(1), 1–19 (1973)
4. A.V. Zayats, I.I. Smolyaninov, A.A. Maradudin, Nano-optics of surface plasmon polaritons. Phys. Rep. **408**, 131–314 (2005)
5. D.K. Gramotnev, S.I. Bozhevolnyi, Plasmonics beyond the diffraction limit. Nat. Photonics **4**, 83–91 (2010)
6. H. Raether, *Surface Plasmons on Smooth and Rough Surfaces and Gratings* (Springer, Berlin, Heidelberg, 1988)
7. S.A. Maier, H.A. Atwater, Plasmonics: Localization and guiding of electromagnetic energy in metal/dielectric structures. J. Appl. Phys. **98**, 011101 (2005)
8. J. Grandidier, G.C.d. Francs, S. Massenot, A. Bouhelier, L. Markey, J.-C. Weeber, C. Finot, A. Dereux, Gain-assisted propagation in a plasmonic waveguide at telecom wavelength. Nano Lett. **9**, 2935–2939 (2009)
9. S.A. Maier, Gain-assisted propagation of electromagnetic energy in subwavelength surface plasmon polariton gap waveguides. Opt. Commun. **258**, 295–299 (2006)
10. M. Nezhad, K. Tetz, Y. Fainman, Gain assisted propagation of surface plasmon polaritons on planar metallic waveguides. Opt. Express **12**, 4072–4079 (2004)
11. M.A. Noginov, G. Zhu, M. Bahoura, J. Adegoke, C.E. Small, B.A. Ritzo, V.P. Drachev, V.M. Shalaev, Enhancement of surface plasmons in an Ag aggregate by optical gain in a dielectric medium. Opt. Lett. **31**, 3022–3024 (2006)
12. Z. Fan, M. Scherbakov, M. Allen, J. Allen, B. Wenner, G. Shvets, Perfect diffraction using all-dielectric bianisotropic metagratings, in *IEEE RAPID*, Miramar Beach, 2018
13. Z. Fan, M.R. Shcherbakov, M. Allen, J. Allen, G. Shvets, "Perfect" diffraction with bianisotropic metagratings. ACS Photonics **5**(11), 4303–4311 (2018)
14. A.L. Fannin, B.R. Wenner, J.W. Allen, M.S. Allen, R. Magnusson, Properties of mixed metal-dielectric nanogratings for application in resonant absorption, sensing, and display. Opt. Eng. **56**(12), 121905 (2017)
15. M. Allen, J.W. Allen, D.M. Wasserman, G.V.P. Kumar, S.A. Maier, Special section guest editorial: Plasmonics systems and applications. Opt. Eng. **56**(12), 121900 (2017)

16. M. Noginov, L. Gu, J. Livenere, G. Zhu, A. Pradhan, R. Mundle, M. Bahoura, Y.A. Barnakov, V. Podolskiy, Transparent conductive oxides: Plasmonic materials for telecom wavelengths. Appl. Phys. Lett. **99**(2), 021101 (2011)
17. G. Naik, J. Kim, N. Kinsey, A. Boltasseva, Alternative plasmonic materials, in *Modern Plasmonics*, ed. By N.V. Richardson, Stephen Holloway (Elsevier Science, North-Holland, 2014), pp. 189–221
18. J.B. Khurgin, Replacing noble metals with alternative materials in plasmonics and metamaterials: How good an idea? Phil. Trans. R. Soc. A **375**, 0068 (2016)
19. M. Allen, J. Allen, T. Schoeppner, D. Look, B. Wenner, Application of highly conductive ZnO to plasmonics, in *GOMACTech*, 2014
20. D.J. Winarski, W. Anwand, A. Wagner, P. Saadatkia, F.A. Selim, M. Allen, B. Wenner, K. Leedy, J. Allen, S. Tetlak, D.C. Look, Induced conductivity in sol-gel ZnO films by passivation or elimination of Zn vacancies. AIP Adv. **6**, 095004 (2016)
21. D. Winarski, W. Anwand, A. Wagner, P. Saadatkia, F. Selim, M. Allen, B. Wenner, K. Leedy, J. Allen, S. Tetlak, D. Look, Induced conductivity in sol-gel ZnO films by passivation or elimination of Zn vacancies, in *18th International Conference on Positron Annihilation*, 2018
22. D. Winarski, J. Ji, F.A. Selim, M. Allen, B.R. Wenner, J. Allen, D.C. Look, Sol-gel synthesis of conductive zinc oxide thin films, in *Electronic Materials Conference*, 2015
23. L. Yang, T. Zhang, H. Zhou, S.C. Price, B.J. Wiley, W. You, Solution-processed flexible polymer solar cells with silver nanowire electrodes. ACS Appl. Mater. Interfaces **3**, 4075–4084 (2011)
24. A.C. Gâlcă, M. Secu, A. Vlad, J.D. Pedarnig, Optical properties of zinc oxide thin films doped with aluminum and lithium. Thin Solid Films **518**, 4603 (2010)
25. K. Ellmer, R. Mientus, Carrier transport in polycrystalline ITO and ZnO:Al II: The influence of grain barriers and boundaries. Thin Solid Films **516**, 5829–5835 (2008)
26. S. Garry, E. McCarthy, J.P. Mosnier, E. McGlynn, Control of ZnO nanowire arrays by nanosphere lithography (NSL) on laser-produced ZnO substrates. Appl. Surf. Sci. **257**, 5159–5162 (2011)
27. M. Ferrera, E.G. Carnemolla, Ultra-fast transient plasmonics using transparent conductive oxides. J. Opt. **024007**, 6 (2018)
28. D.F. Liu, Y.J. Xiang, X.C. Wu, Z.X. Zhang, L.F. Liu, L. Song, X.W. Zhao, S.D. Luo, W.J. Ma, J. Shen, W.Y. Zhou, G. Wang, S.S. Xie, Periodic ZnO nanorod arrays defined by polystyrene microsphere self-assembled monolayers. Nano Lett. **6**, 2375–2378 (2006)
29. K.F. MacDonald, Z.L. Samson, M.I. Stockman, N.I. Zheludev, Ultrafast active plasmonics. Nat. Photonics **3**, 55–58 (2009)
30. H.T. Ng, B. Chen, J. Li, J. Han, M. Mayyappan, J. Wu, S.X. Li, E.E. Haller, Optical properties of single-crystalline ZnO nanowires on m-sapphire. Appl. Phys. Lett. **2023–2025**, 82 (2003)
31. K. Tominaga, H. Manabe, N. Umezu, I. Mori, T. Ushiro, I. Nakabayashi, Film properties of ZnO: Al prepared by cosputtering of ZnO:Al and either Zn or Al targets. J. Vac. Sci. Technol. A **15**, 1074–1079 (1997)
32. H. Kim, J.S. Horwitz, S.B. Qadri, D.B. Chrisey, Epitaxial growth of Al-doped ZnO thin films grown by pulsed laser deposition. Thin Solid Films **420–421**, 107–111 (2002)
33. S. Dev, D. Look, K. Leedy, L. Yu, D. Walker Jr., B. Wenner, J. Allen, M. Allen, D. Wasserman, Gallium-doped zinc oxide plasmonic nanostructures for mid-IR applications, in *SPIE Optics+Photonics*, San Diego, 2016
34. A. Suzuki, M. Nakamura, R. Michihata, T. Aoki, T. Matsushita, M. Okuda, Ultrathin Al-doped transparent conducting zinc oxide films fabricated by pulsed laser deposition. Thin Solid Films **517**, 1478–1481 (2008)
35. J. Allen, M. Allen, D. Look, B. Wenner, N. Itagaki, K. Matsushima, I. Surhariadi, Infrared plasmonics via ZnO. J. Nano Res. **28**, 109 (2014)
36. M.S. Allen, J.W. Allen, B.R. Wenner, D.C. Look, K.D. Leedy, Application of highly conductive ZnO to plasmonics, in *Proc. SPIE 8626, Oxide-Based Materials and Devices IV, 862605*, 2016

37. T. Minami, T. Miyata, Y. Ohtani, K. Kuboi, Effect of thickness on the stability of transparent conducting impurity-doped ZnO thin films in a high humidity environment. Phys. Status Solidi RRL **1**, R31–R33 (2007)
38. D.C. Look, K.D. Leedy, D.H. Tomich, B. Bayraktaroglu, Mobility analysis of highly conducting thin films: Application to ZnO. Appl. Phys. Lett. **96**, 062102–062103 (2010)
39. M. Allen, J. Allen, B. Wenner, D. Look, K. Leedy, Application of highly conductive ZnO to the excitation of long-range plasmons in symmetric hybrid waveguides. Opt. Eng. **52**(6), 064603–064603 (2013)
40. D.C. Look, K.D. Leedy, A. Kiefer, B. Claflin, N. Itagaki, K. Matsushima, I. Surhariadi, Model for thickness dependence of mobility and concentration in highly conductive zinc oxide. Opt. Eng. **52**, 033801–033801 (2013)
41. N. Itagaki, K. Kuwahara, K. Matsushima, K. Oshikawa, Novel fabrication method for ZnO films via nitrogen-mediated crystallization, in *SPIE OPTO*, San Francisco, 2012
42. N. Itagaki, K. Kuwahara, K. Nakahara, D. Yamashita, G. Uchida, K. Koga, M. Shiratani, Highly conducting and very thin ZnO:Al films with ZnO buffer layer fabricated by solid phase crystallization from amorphous phase. Appl. Phys. Express **4**, 011101 (2011)
43. K. Kuwahara, N. Itagaki, K. Nakahara, D. Yamashita, G. Uchida, K. Kamataki, K. Koga, M. Shiratani, High quality epitaxial ZnO films grown on solid-phase crystallized buffer layers. Thin Solid Films **520**, 4674–4677 (2012)
44. D.C. Look, T.C. Droubay, S.A. Chambers, Optical/electrical correlations in ZnO: The plasmonic resonance phase diagram. Phys. Status Solidi B **250**(10), 2118–2121 (2013)
45. D.C. Look, Two-layer Hall-effect model with arbitrary surface-donor profiles: Application to ZnO. J. Appl. Phys. **104**, 063718–063717 (2008)
46. D. Look, J. Allen, M. Allen, B. Wenner, N. Itagaki, K. Matsushima, I. Surhariadi, Infrared plasmonics via ZnO, in *4th Mexican Workshop on Nanostructured Materials*, 2013
47. D. Look, M. Allen, J. Allen, B. Wenner, N. Itagaki, K. Matsushima, I. Surhariadi, Infrared plasmonics via ZnO, in *Joint Symposia of the Japanese Society of Applied Physics and International Materials Research Society*, 2013
48. D.C. Look, K.D. Leedy, ZnO plasmonics for telecommunications. Appl. Phys. Lett. **102**, 182104–182107 (2013)
49. T. Schumann, J. Neff, S. Breedlove, H. Zmuda, Y.-K. Yoon, D. Look, K. Leedy, M. Allen, J. Allen, Optical transmittance and reflectance of lanthanum nickelate at telecommunication frequencies, in *IEEE RAPID*, Miramar Beach, 2018
50. T. Paik, S. Hong, E. Gaulding, H. Caglayan, T.R. Gordon, N. Engheta, C. Kagan, C. Murray, Solution processed phase change VO2 metamaterials from colloidal vanadium oxide nanocrystals. ACS Nano **8**(1), 797–806 (2014)
51. C. Metaxa, S. Kassavetis, J. Pierson, D. Gall, P. Patsalas, Infrared plasmonics with conductive ternary nitrides. ACS Appl. Mater. Interfaces **9**, 10825–10834 (2017)
52. U. Guler, G.V. Naik, A. Boltasseva, V.M. Shalaev, A.V. Kildishev, Nitrides as alternative materials for localized surface plasmon applications, in *FiO/LS Technical Digest*, 2012
53. G.V. Naik, J. Schroeder, U. Guler, X. Ni, A. V. Kildishev, T.D. Sands, A. Boltasseva, Metal nitrides for plasmonic applications, in *CLEO Technical Digest*, 2012
54. R.E. Simpson, J. Renger, M. Rude, R. Quidant, V. Pruneri, Active plasmonics based on phase change materials, in *European Conference on Integrated Optics*, 2016
55. P. Patsalas, N. Kalfagiannis, S. Kassavetis, G. Abadias, D. Bellas, C. Lekka, E. Lidorikis, Conductive nitrides: Growth principles, optical and electronic properties, and their perspectives in photonics and plasmonics. Mater. Sci. Eng. R **123**, 1–55 (2018)
56. H. Reddy, D. Shah, N. Kinsey, V.M. Shalaev, A. Boltasseva, Ultra-thin plasmonic metal nitrides: Tailoring optical properties to photonic applications, in *International Conference on Optical MEMS and Nanophotonics*, Santa Fe, 2017
57. K.K. Madapu, A.K. Sivadasan, M. Baral, S. Dhara, Observation of surface plasmon polaritons in 2D electron gas of surface electron accumulation in InN nanostructures. Nanotechnology **29**, 275707 (2018)

58. G.A. Melentev, D.Y. Yaichnikov, V.A. Shalygin, M.Y. Vinnichenko, L.E. Vorobjev, D.A. Firsov, L. Riuttanen, S. Suihkonen, Plasmon phonon modes and optical resonances in n-GaN. J. Phys. Conf. Ser. **690**, 012005 (2016)

59. G.A. Melentev, V.A. Shalygin, L.E. Vorobjev, V.Y. Panevin, D.A. Firsov, L. Riuttanen, S. Suihkonen, V.V. Korotyeyev, Y.M. Lyaschuk, V.A. Kochelap, V.N. Poroshin, Interaction of surface plasmon polaritons in heavily doped GaN microstructures with terahertz radiation. J. Appl. Phys. **119**(9), 093104 (2016)

60. A. Boltasseva, Unlocking new physics and enabling plasmonic and metamaterial devices with improved materials, in *DTIC*, 2014

61. A. Rosenberg, J. Surya, R. Liu, W. Streyer, S. Law, L.S. Leslie, R. Bhargava, D. Wasserman, Flat mid-infrared composite plasmonic materials using lateral doping-patterned semiconductors. J. Opt. **16**, 094012 (2014)

62. Y. Zhong, S.D. Malagari, T. Hamilton, D. Wasserman, Review of mid-infrared plasmonic materials. J. Nanophotonics **9**, 093791–093791 (2015)

63. S. Law, D.C. Adams, A.M. Taylor, D. Wasserman, Mid-infrared designer metals. Opt. Express **20**(11), 12155–12165 (2012)

64. A.J. Hoffman, L. Alekseyev, S.S. Howard, K.J. Franz, D.M. Wasserman, V.A. Podolskiy, E.E. Narimanov, D.L. Sivco, C. Gmachl, Negative refraction in semiconductor. Nat. Mater. **6**, 946 (2007)

65. M.P. Fischer, A. Riede, A. Grupp, K. Gallacher, J. Frigerio, M. Ortolani, D.J. Paul, G. Isella, A. Leitenstorfer, P. Biagioni, D. Brida, Germanium plasmonic nanoantennas for third-harmonic generation in the mid infrared, in *2016 41st International Conference on Infrared, Millimeter, and Terahertz waves (IRMMW-THz)*, 2016

66. J.A. Dionne, L.A. Sweatlock, M.T. Sheldon, A.P. Alivisatos, H.A. Atwater, Silicon-based plasmonics for on-chip photonics. IEEE J. Sel. Top. Quantum Electron. **16**(1), 295 (2010)

67. Q. Guo, C. Li, B. Deng, S. Yuan, F. Guinea, F. Xia, Infrared nanophotonics based on graphene plasmonics. ACS Photonics **4**, 2989–2999 (2017)

68. S. Xiao, X. Zhu, B. Li, Graphene-plasmon polaritons: From fundamental properties to potential applications. Front. Phys. **11**(2), 117801 (2016)

69. B.D. Fainberg, G. Li, Nonlinear organic plasmonics: Applications to optical control of Coulomb blocking in nanojunctions. Appl. Phys. Lett. **107**, 053302 (2015)

70. B.D. Fainberg, N.N. Rosanov, N.A. Veretenov, Light-induced "plasmonic" properties of organic materials: Surface polaritons and switching waves in bistable organic thin films. Appl. Phys. Lett. **110**, 203301 (2017)

71. W. Barnes, Organic materials instead of metals for plasmonics, in *2017 Conference on Lasers and Electro-Optics Europe & European Quantum Electronics Conference (CLEO/Europe-EQEC)*, Munich, 2017

72. J. Seidel, S. Grafström, L. Eng, Stimulated emission of surface plasmons at the interface between a silver film and an optically pumped dye solution. Phys. Rev. Lett. **94**, 177401 (2005)

73. L. Gu, J. Livenere, G. Zhu, E.E. Narimanov, M.A. Noginov, Quest for organic plasmonics. Appl. Phys. Lett. **103**, 021104 (2013)

74. S. Trendafilov, M. Allen, J. Allen, Y.J. Yoon, Y. Chen, C.R. Gomez, Z. Lin, To etch or not to etch, in *IEEE RAPID*, Miramar Beach, 2018

75. A.M. Alkilany, C.J. Murphy, Toxicity and cellular uptake of gold nanoparticles: What we have learned so far? J. Nanopart. Res. **12**, 2313–2333 (2010)

76. V. Amendola, R. Pilot, M. Frasconi, O.M. Maragò, M.A. Iatì, Surface plasmon resonance in gold nanoparticles: A review. J. Phys. Condens. Matter **29**, 203002 (2017)

77. I. Pastoriza-Santos, C. Kinnear, J. Pérez-Juste, P. Mulvaney, L.M. Liz-Marzán, Plasmonic polymer nanocomposites. Nat. Rev. Mater. **3**, 375 (2018)

78. W.T. Hsieh, P.C. Wu, J.B. Khurgin, D.P. Tsai, N. Liu, G. Sun, Comparative analysis of metals and alternative infrared plasmonic materials. ACS Photonics **8**, 2541–2548 (2017)

79. Z.V. Vardeny, T. Matsui, A. Agrawal, A. Nahata, R. Menon, Tunable plasmonic crystals of heavily-doped conducting polymers promise novel devices, in *SPIE Newsroom,* 1 April 2006
80. A.X. Wang, X. Kong, Review of recent progress of plasmonic materials and nano-structures for surface-enhanced Raman scattering. Materials **8**, 3024–3052 (2015)
81. B. Grillo, M. Allen, B. Wenner, J. Allen, C. Murphy, Switchable optical materials through alignment of gold nanorods, in *GOMACTech*, 2014
82. J.S. Biggins, S. Yazdi, E. Ringe, Magnesium nanoparticle plasmonics. Nano **18**, 3752–3758 (2018)
83. H. Matsui, H. Tabata, Oxide semiconductor nanoparticles for infrared plasmonic applications, in *Science and Applications of Tailored Nanostructures*, One Central Press, Altrincham, (2012), pp. 68–86)
84. D.B. Tice, S.Q. Li, M. Tagliazucchi, D.B. Buchholz, E.A. Weiss, R.P.H. Chang, Ultrafast modulation of the plasma frequency of vertically aligned indium tin oxide rods. Nano **14**(3), 1120–1126 (2014)
85. A.M. Schimpf, S.D. Lounis, E.L. Runnerstrom, D.J. Milliron, D.R. Gamelin, Redox chemistries and plasmon energies of photodoped In_2O_3 and Sn-doped In_2O_3(ITO) nanocrystals. J. Am. Chem. Soc. **137**, 518–524 (2015)
86. R. Buonsanti, A. Llordes, S. Aloni, B.A. Helm, D.J. Milliron, Tunable infrared absorption and visible transparency of colloidal aluminum-doped zinc oxide nanocrystals. Nano **11**(11), 4706–4710 (2011)
87. I. Kriegel, J. Rodríguez-Fernandez, A. Wisnet, H. Zhang, C. Waurisch, A. Eychmüller, A. Dubavik, A.O. Govorov, J. Feldmann, Shedding light on vacancy-doped copper chalcogenides: Shape-controlled. ACS Nano **7**, 4367–4377 (2013)
88. H. Zhang, R. Zhang, K.S. Schramke, N.M. Bedford, K. Hunter, U.R. Kortshagen, P. Nordlander, Doped silicon nanocrystal plasmonics. ACS Photonics **4**, 963–970 (2017)
89. B. Khlebtsov, E. Panfilova, V. Khanadeev, O. Bibikova, G. Terentyuk, A. Ivanov, V. Rumyantseva, I. Shilov, A. Ryabova, V. Loshchenov, N.G. Khlebtsov, Nanocomposites containing silica-coated gold–silver nanocages and Yb–2,4-Dimethoxyhematoporphyrin: Multifunctional capability of IR-luminescence detection, photosensitization, and photothermolysis. ACS Nano **5**(9), 7077–7089 (2011)
90. D. Yang, X. Pang, Y. He, Y. Wang, G. Chen, W. Wang, Z. Lin, Precisely size-tunable magnetic/plasmonic core/shell nanoparticles with controlled optical properties. Angew. Chem. **54**, 12091–12096 (2015)
91. Y.J.Y. Chen, C.R. Gomez, M. Allen, J. Allen, Z. Lin, Synthesis and characterizations of plasmonic nanoparticles: Large plain Au and Au/TiO_2 core-shell nanoparticles, in *IEEE RAPID*, Miramar Beach, 2018
92. C.S. Levin, C. Hofmann, T.A. Ali, A.T. Kelly, E. Morosan, P. Nordlander, K.H. Whitmire, N.J. Halas, Magnetic-plasmonic core-shell nanoparticles. ACS Nano **3**(6), 1379–1388 (2009)
93. C.P. Byers, H. Zhang, D.F. Swearer, M. Yorulmaz, B.S. Hoener, D. Huang, A. Hoggard, W.-S. Chang, P. Mulvaney, E. Ringe, N.J. Halas, P. Nordlander, S. Link, C.F. Landes, From tunable core-shell nanoparticles to plasmonic drawbridges: Active control of nanoparticle optical properties. Sci. Adv. **1**(11), 1500988 (2015)
94. M.W. Knight, N.J. Halas, Nanoshells to nanoeggs to nanocups: Optical properties of reduced symmetry core–shell nanoparticles beyond the quasistatic limit. New J. Phys. **10**, 105006 (2008)
95. S.D. Campbell, R.W. Ziolkowski, Simultaneous excitation of electric and magnetic dipole modes in a resonant core-shell particle at infrared frequencies to achieve minimal backscattering. IEEE J. Sel. Top. Quantum Electron. **19**(3), 4700209 (2013)
96. S.H. Behrens, V. Breedveld, M. Mujica, M.A. Filler, Process principles for large-scale nanomanufacturing. Annu. Rev. Chem. Biomol. **8**, 201–226 (2017)
97. S. Braswell, B. Wenner, M. Allen, J. Allen, A. Ephrem, L.L. Goddard, A computational study of a hybrid plasmonic-microring for label-free detection, in *IEEE Photonics Conference*, Hawaii, 2016

98. K.J. Lee, R. Magnusson, B.R. Wenner, J.W. Allen, M. S. Allen, Neuropeptide Y binding dynamics quantified with nanophotonic resonant sensors, in *IEEE RAPID*, Miramar Beach, 2018
99. M.I. Stockman, K. Kneipp, S.I. Bozhevolnyi, S. Saha, A. Dutta, J. Ndukaife, N. Kinsey, H. Reddy, U. Guler, V.M. Shalaev, A. Boltasseva, B. Gholipour, H.N.S. Krishnamoorthy, K. MacDonald, Roadmap on plasmonics. J. Opt. **043001**, 39 (2018)
100. T. Low, P. Avouris, Graphene plasmonics for terahertz. ACS Nano **8**(2), 1086–1101 (2014)
101. R. Stanley, Plasmonics in the mid-infrared. Nat. Photonics **6**, 409–412 (2012)
102. A.J. Rahedi, J.F. Douglas, F.W. Starr, Model for reversible nanoparticle assembly in a polymer matrix. J. Chem. Phys. **128**(2), 024902 (2008)
103. S. Cho, M.A. Ciappesoni, M.S. Allen, J.W. Allen, K.D. Leedy, B.R. Wenner, S.J. Kim, Efficient broadband energy detection from the visible to near-infrared using a plasmon FET. Nanotechnology **29**(28), 285201 (2018)

Chapter 6
Materials for Flexible Thin-Film Transistors: High-Power Impulse Magnetron Sputtering of Zinc Oxide

Amber N. Reed

6.1 Introduction

Zinc oxide is a wide band gap (3.3–3.6 eV) semiconductor that has been the focus of considerable research [1–6] for use in electronic applications due to its promising electrical transport properties without the need for epitaxial growth. Field-effect mobilities (μ_F) up to 110 cm^2/(V s) have been reported for polycrystalline ZnO [7, 8]. These values approach the mobilities reported for single-crystalline Si ($\mu_F \sim$ 200 cm^2 V^{-1} s^{-1}) [9] and surpass the mobilities of less than 5 cm^2/(V s) reported for amorphous silicon and organic semiconductors [9]. The high mobility of polycrystalline ZnO is attributed to the material's ionic bonding [10]. Unlike in covalently bonded semiconductors, such as Si, overlap of ZnO's oxygen conduction band 2s orbitals with its neighboring Zn 4s orbitals provides paths for electrons that are unaffected by distortions in the Zn-O chemical bonds [10]. A similar effect is observed in amorphous oxide semiconductors (such as indium gallium zinc oxide) [10]. In addition to its high mobility, ZnO possesses a high breakdown voltage, the ability to sustain large electric fields, low noise generation, and high-temperature high-power operations due to its wide band gap [1]. High temperature stability and radiation hardness [11, 12] allow ZnO to tolerate extreme environments. Additionally, its wide band gap makes ZnO optically transparent allowing for its incorporation into transparent electronics. These properties make ZnO attractive as a channel material for a variety of optoelectronic and thin-film field-effect transistor (TFT) devices used in active displays, solar cells, chemo- and bio-sensors, RF signal processing, power electronics, and other applications [1–3].

The on/off ratio, peak current density, and saturation current of the device will dictate the TFT performance. These parameters are affected by the channel material

A. N. Reed (✉)
Materials and Manufacturing Directorate, Air Force Research Laboratory,
Wright-Patterson Air Force Base, OH, USA
e-mail: amber.reed.5@us.af.mil

© Springer Nature Switzerland AG 2020　　　　　　　　　　　　　　　　79
M. E. Kinsella (ed.), *Women in Aerospace Materials*, Women in Engineering
and Science, https://doi.org/10.1007/978-3-030-40779-7_6

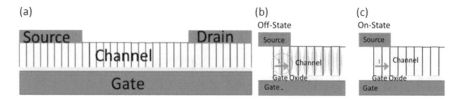

Fig. 6.1 (**a**) Schematic diagram of TFT with a ZnO channel. (**b**) Close-up of the TFT channel showing the depletion zones at the grain boundaries in the off-state and (**c**) the collapse of the depletion zones under the gate field in the on-state

in the TFT. High-performance TFTs, such as microwave FETs, require large current densities and high on/off ratios. The pulsed laser deposition (PLD) of ZnO channels for microwave FET has been previously demonstrated by Bayraktaroglu, Leedy, and Neidhard [7]. In their work, Bayraktaroglu and Leedy attribute the high device performance in their ZnO films to the material's continuous closely packed nanocolumnar structure [8]. The high on/off ratios that are reported are possible due to the closely packed vertical grain boundaries creating back-to-back depletion zones that prevent horizontal current flow from the source to drain when the TFT is in the off-state (as illustrated in Fig. 6.1). When a potential is applied to the gate during the on-state, the depletion zones collapse, and current is able to flow through the channel. The dense nanocolumnar structure of the ZnO films is not disrupted when the film passes over non-planar surfaces, for example, the transition from SiO_2 to the raised metal gate [8]. Because the conduction mechanism in ZnO is controlled by point defects (i.e., interstitial zinc atoms and oxygen vacancies) in the film, control of the film crystallinity is crucial for TFT applications [6, 12–14]. The choice of deposition technique and conditions can play a critical role in determining ZnO crystallinity.

In addition to PLD [15–21], ZnO films have been deposited using chemical vapor deposition [22, 23], atomic layer deposition [24], solution-based hydrothermal deposition [25], filtered vacuum cathodic arc deposition [26–28], radio frequency (RF) sputtering [14, 29–32], DC and pulsed DC magnetron sputtering [5, 33], and high-power impulse magnetron sputtering (HiPIMS) [34–36]. These techniques all result in wurtzite ZnO with a (002) preferred orientation. The ZnO crystal and microstructural properties, such as mosaicity, presence of other crystallographic orientations, grain size, and surface roughness of the films, however, are affected by the energy of the film material on the substrate. The processing technique, background pressure, and substrate temperature used during film growth determine these energies. Low energy (and low temperatures) at the substrate results in low surface mobility for the growing film and causes the formation of defects and many small grains. This can result in lower film conductivity or mobility. Higher substrate temperatures, higher powers during deposition, or using techniques with inherently higher energy fluxes, like RF sputtering or HiPIMS, can be used to increase the energy of the film species.

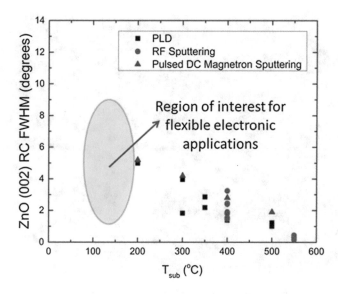

Fig. 6.2 Summary of FWHM of (002) rocking curve XRD peaks at different substrate temperatures for ZnO films deposited with PLD [27–29], RF sputtering [31, 32], and pulsed DC magnetron sputtering [33]

A common technique to determine crystal quality is x-ray diffraction (XRD). The width of an XRD rocking curve is a particularly useful indicator of crystal quality as it can be used to determine the orientation of the crystal. The effect of substrate temperature and processing technique on film crystallinity, specifically the degree of alignment of the (002) orientation relative to the substrate, for films synthesized with physical vapor deposition (PVD) techniques is illustrated in Fig. 6.2. The graph shows the full width half maximum (FWHM) of the ZnO (002) rocking curve peak as a function of deposition temperature. The highest degree of (002) alignment, indicated by the smallest FWHM, occurs for films deposited with RF magnetron sputtering or high substrate temperatures. Those deposited with lower temperatures have a broader mosaic spread. There is currently little information on the orientation of PVD synthesized ZnO with substrate temperatures below 200 °C. This is a critical processing region for incorporation of ZnO films into flexible electronic applications and the focus of this study.

Previous studies on PLD of ZnO have shown that high carrier mobility and large on/off ratios require films with low surface roughness, dense morphology, and strong (002) orientation [7, 8]. There are, however, intrinsic challenges to the scalability of PLD thin-film growth for large areas and complex-shaped substrates. These challenges motivate investigation of low-temperature and scalable deposition techniques to yield the material morphology and structure optimization identified from the PLD growth studies. Magnetron sputtering techniques have a great deal of potential for scalability of ZnO synthesis at low temperatures due to the technique's energetic plasma fluxes. However, initial investigations with conventional

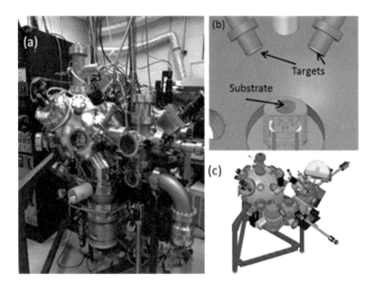

Fig. 6.3 (**a**) Photograph of the vacuum chamber used in the experiments. Schematic of the (**b**) interior and (**c**) exterior of the deposition chamber

magnetron sputtering reported in literature show ZnO films with low mobilities ($\mu_{FE} \leq 1.8$ cm^2 V^{-1} s^{-1}) [37] and small on/off ratios, due to poor crystallinity, and an open columnar structure with a high degree of porosity. RF magnetron sputtering, which produces smooth crystalline films with highly aligned (002) crystals, is difficult to scale to industrial size due to the need for impedance matching of the substrate-plasma system. HiPIMS is a highly scalable technique with high ion fluxes and ion energies [38, 39] having great potential as a technique for large area growth of ZnO with controlled microstructures and crystallographic orientation.

Konstantinidis, Hemberg, Dauchot, and Hecq have previously demonstrated reactive HiPIMS of ZnO films on glass [34], while Partridge, Mayes, McDougall, Bilek, and McCulloch have demonstrated it on sapphire substrates [35]. In both studies, the ZnO films were nanocrystalline with small 1200–1300 nm^2 wurtzite grains and preferred c-axis orientation. The ZnO on sapphire had Hall mobilities of 7–8 cm^2 V^{-1} s^{-1}, which are comparable to those reported for DC magnetron sputtering [40] but significantly lower than the Hall mobilities of >50 cm^2 V^{-1} s^{-1} reported for ZnO films grown with PLD [7, 8].

6.2   Experimental Details

In this study, plasma characterization and film growth were performed in the custom-built ultra-high vacuum system (shown in Fig. 6.3). A Huettinger HMP-1200 HiPIMS power supply was used to apply pulses of voltage to a 3.30 cm Zn target. Before any ZnO films were grown, a thorough study of the effect of deposition conditions on plasma stability and target current (I_T) was completed. A Tektronix TM502A current probe and Tektronix TDS 5400 oscilloscope were used to measure I_T. A Tektronix P5200 high-voltage differential probe was used to trigger the current measurements on the oscilloscope to ensure that the recorded currents were time-resolved and allowed the evolution of the current to be correlated to pulse initiation and termination. From initial studies of the plasma, a total pressure of 2.67 Pa and an O_2/Ar of 0.4 were chosen. The target current was then measured while systematically varying applied voltage (V_T) from 300 V to 650 V, while pulse length (τ) and pulse frequency (f) were held constant at 200 µs and 100 Hz, respectively. A second set of I_T measurements were performed with $V_T = 500$ V, $f = 100$ Hz, and τ varied from 20 µs to 200 µs. The third set of I_T measurements varied f from 10 Hz to 200 Hz while $V_T = 500$ V and $\tau = 200$ µs.

The results from the I_T measurements were then used to select a range of conditions with which to grow ZnO films to determine the effect of V_T, τ, and f on ZnO crystallinity. Three series of 100-nm-thick ZnO films were prepared on 1 cm^2 conductive silicon (Si) substrates with a 100-nm-thick thermal silicon oxide (SiO_2) layer. The substrates were cleaned using the sonicating and rinsing procedure described in reference [41] and masked with a small piece of Kapton tape to allow ex situ measurement of film thickness. The substrates where then mounted to stainless steel discs using double-sided carbon tape and loaded into a small antechamber. The antechamber, which was capable of holding five samples, was evacuated to pressures below 10^{-4} Pa before the samples were individually moved to the deposition chamber. The deposition chamber had a base pressure below 10^{-6} Pa. All nine films were prepared in an O_2/Ar gas mixture with a total pressure of 2.67 Pa and an oxygen partial pressure of 1.07 Pa. All films were prepared with the electrically grounded substrate held at 200 °C. The substrates were rotated to promote film uniformity. For the set of films investigating the effects of target voltage on film growth, three ZnO films were prepared with $V_T = 350$ V, 500 V, and 600 V, while τ and f were held constant at 200 µs and 100 Hz, respectively. For the second set of films, the V_T and f were held constant (500 V and 100 Hz), while τ was changed ($\tau = 50$ µs, 100 µs, 200 µs). The final set of films were grown with different pulse frequencies ($f = 50$ Hz, 100 Hz, 250 Hz), while V_T and τ were fixed at 500 V and 200 µs. A small section of the substrates was masked with Kapton tape to allow ex situ measurement of the film thickness. The substrates were electrically grounded and heated to 150 °C.

The ZnO film crystallinity and orientation were analyzed with XRD using a Rigaku Smartlab XRD system with a Cu anode. The beam divergence was minimized using a 1 mm divergence slit and a 0.5° parallel slit analyzer. Parallel beam $2\Theta/\omega$ wide-angle XRD scans were taken from $2\Theta = 20$–60°, with 0.002° steps at scan rate of 0.2°/minute, to determine the crystal orientations present in the films. The XRD patterns were fitted using a Pearson VII shape function to determine peak FWHM and integrated intensity.

After the films were grown, the substrate holder was removed and replaced with an energy-resolved Hiden EQP 1000 electrostatic quadrupole mass spectrometer. The mass and ion energy distributions (IEDs) of the flux arriving at the substrate

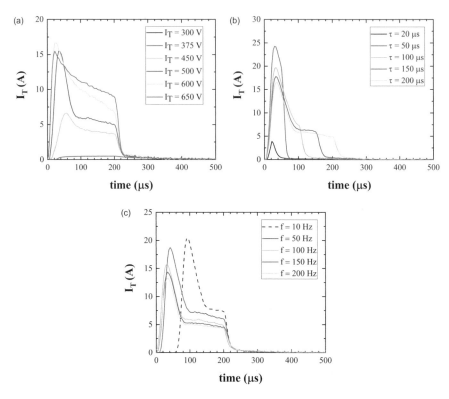

Fig. 6.4 Time-resolved current of zinc target with different (**a**) applied voltage, (**b**) pulse length, and (**c**) pulse frequency

were measured for each of the deposition conditions. IEDs from 0 eV to 90 eV with steps of 0.50 eV were measured for mass-charge ratios of 16 amu/C (O+), 32 amu/C (O$_2$+), 40 amu/C (Ar+), 65 amu/C (Zn+), and 80 amu/C (ZnO+) at the same conditions as the mass scans. A dwell time of 300 ms was used for both the mass scans and IEDs to ensure each data point for the standard pulse conditions (200 μs; 100 Hz) was composed of 30 pulses.

6.3 Results and Discussion

The time-resolved currents for the Zn target (Fig. 6.4) showed a strong dependence on applied voltage and pulse parameters. The dependence of target current on applied voltage can be seen in Fig. 6.4a. Applied voltages lower than 450 V resulted in an unstable plasma and negligible measured current (<1 mA). For applied voltages above 450 V, the general character of the current shape remained constant. The current waveforms consisted of three characteristic regions corresponding to plasma initiation, steady state, and pulse-off times. At pulse initiation, the current rose. As the zinc oxide layer on the target surface was bombarded with argon ions and electrons, oxygen atoms/ions and zinc atoms/ions were ejected. The ejected electrons and ions contributed to the current, and the current increased until a peak value was reached. As the density of sputtered atoms near the target increased, gas rarefaction occurred. Gas rarefaction resulted in a reduction in the flux of argon ions at the target surface [42, 43] and caused a decrease in the target current after the initial peak. The current decayed until a steady-state current was reached. At pulse termination, the current quickly decayed to zero. The magnitude of the peak and the steady-state currents were voltage-dependent. When the target voltage was increased from 450 V to 500 V, the peak current increased from 6.6 A to 15.5 A. Also, the peak value was reached 24 μs earlier. The increase in peak current was less dramatic when the voltage was increased above 500 V; however, the time required to reach the maximum current value continued to decrease at higher voltages.

Both of the other two deposition parameters investigated in this study, t and f, dictate the time available for processes, such as ZnO layer formation on the target surface, sputtering of the ZnO layer, and gas rarefaction, to occur. The formation and removal (via sputtering) of the ZnO layer that forms on the Zn target surface has significant impact on the magnitude of I_T. This is due to the higher secondary electron yield (γ_{SE}) of ZnO compared to Zn. The peak value for I_T was strongly affected by τ, as seen in Fig. 6.4b. The shortest pulse length capable of sustaining a stable plasma was $\tau = 20$ μs; however, I_T for this pulse length was significantly lower than that for longer pulses. The low peak current for the 20 μs pulse can be attributed to the short time for generation of ions and electrons by sputtering before pulse termination. For both the 20 μs and 50 μs long pulses, pulse termination occurred before the current reached steady state. The largest peak current ($I_p = 24$ A) occurred during the 50 μs pulse. The peak current decreased as pulse length was increased. This decrease was likely due to the longer pulses removing more of the ZnO layer that had formed on the target surface between pulses.

Similar behavior was observed for pulse frequency, as seen in Fig. 6.4c. The time between pulses is particularly important during reactive sputtering because it affects how large of a poisoned layer can develop between pulses. At low pulse frequencies ($f \leq 40$ Hz), the plasma was unstable and exhibited a strobe-like pulsing. The current profile for these pulses was unstable, and the current profile shifted between beginning at pulse initiation and rising after a 50 μs delay. Above 40 Hz, I_T was stable and rose quickly to a peak value. The current profile at 50 Hz had a higher peak and steady-state current. At higher pulse frequencies, the general shape and time evolution of the current profile remained the same; however, the peak current value decreased as the time between pulses was increased. A shorter time between pulses with higher pulse frequency allowed less time for oxide formation on the target surface resulting in a reduction in the current.

A range of operating parameters for film growth from the Zn target was determined using the I_T measurements and observations on plasma stability and changes in plasma intensity and emission color. The deposition rate for the films was calculated from the measured film thickness and deposition time. The deposition rate increased from 5.37 ± 0.33 nm/min at 450 V to 27.73 ± 0.30 nm/min at 600 V. The increase in deposition rate with higher applied voltages was much larger than the expected increase in sputtering yield. The calculated deposition rate was higher for films grown at higher pulse lengths, 4.10 ± 0.33 nm/min for 50 μs pulses and 15.47 ± 0.30 nm/min for 200 μs pulses. The deposition rate also increased with high pulse frequencies, 4.93 ± 0.73 nm/min for 50 Hz and 28.07 ± 0.70 nm/min for 250 Hz.

The $2\Theta/\omega$ and rocking curve XRD patterns for these films are shown in Fig. 6.5. All growths in this study resulted in wurtzite ZnO with a preferred (002) orientation. The crystalline quality and ZnO orientation were strongly affected by the target voltage and pulse parameters during growth. The applied target voltage had the greatest effect on ZnO crystallinity. The film grown with the lowest voltage (i.e., 350 V) showed a single low-intensity diffraction peak for $2\Theta = 34.5°$. An order of magnitude increase in the (002) peak intensity for the $2\Theta/\omega$ scan was achieved when the target voltage was increased. The ZnO (002) diffraction peak remained the prominent feature for the films grown with applied voltages of 500 V and 600 V; however, both films had other orientations present. The XRD pattern for the ZnO films grown with 500 V showed diffraction peaks for (100)- and (101)-orientated ZnO, while that for the ZnO grown with 600 V had only diffraction peaks for the (002) and (101) orientations. The diffraction peak for the silicon substrate is also visible in the film grown at 600 V. The rocking curve of these films (Fig. 6.5d) showed broad and low-intensity ω peaks for $V_T = 350$ V and 500 V, indicating that the films grown were highly misoriented.

Figure 6.5b shows the $2\Theta/\omega$ XRD scan of films grown with 50 μs, 100 μs, and 200 μs pulses. The pulse lengths selected for film growth encompass the three target current regimes observed in Fig. 6.5b: (1) where the pulse terminates immediately after the peak current is reached; (2) where the pulse terminates after the current has decayed but before a steady-state value is reached; and (3) where the pulse terminates after the steady state has been reached. The ZnO (002) diffraction peak is the prominent feature in all the XRD patterns. For the 50 μs film, the (002) peak is the only feature. The wide-angle XRD patterns for film grown with different pulse frequencies are compiled in Fig. 6.5c. No shift in the (002) diffraction peak was observed with changes in the pulse frequency. The concentration of (101) and (100) orientations changed as higher pulse frequencies reduced the time between pulses for the deposited material to reach energetically favorable orientations before burial by the subsequent incoming fluxes. The rocking curve of these films (Fig. 6.5e) showed that misorientation decreased for films grown with higher pulse lengths.

The IEDs for Ar^+, O^+, O_2^+, Zn^+, and ZnO^+ for different applied target currents are compiled in Fig. 6.6. For Ar^+ at voltages of 450 V or lower, the plasma was weak, and there were few energetic ions. At 500 V and greater, a strong single peak with an average energy of 4 eV was measured. The IED for 700 V had peaks at 3 eV and 5.5 eV. The lower-energy ion was attributed to ions that have lost energy due to collisions with thermalized gas atoms and ions [44]. The intensity of the lower-energy peak was lower at 700 V than at 600 V. The reduction in counts at 700 V was likely

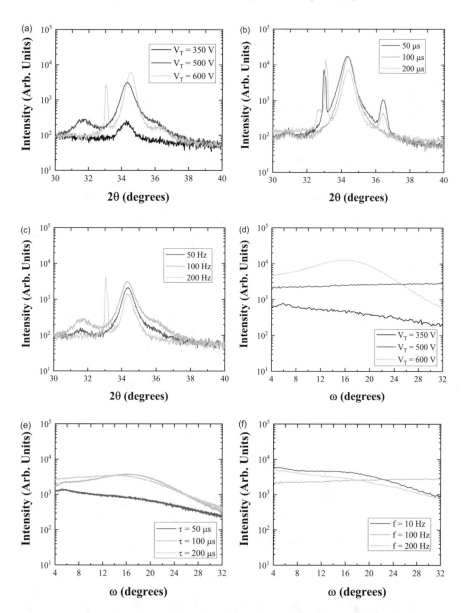

Fig. 6.5 Wide-angle XRD patterns for ZnO films deposited with different (**a**) applied target voltages, (**b**) pulse lengths, and (**c**) pulse frequencies and rocking curve XRD for ZnO films deposited with different (**d**) applied target voltages, (**e**) pulse lengths, and (**f**) pulse frequencies

a result of the thicker plasma sheath ($s = 0.28$ cm and 700 V vs. $s = 0.26$ cm at 600 V) increasing the probability of Ar^+ experiencing energy-reducing collisions.

The IED for O^+ in Fig. 6.6b showed two peaks with average energies of 2 eV and 5 eV. At 500 V, the IED had an intensity of 5.0×10^4 counts/second at 1 eV and

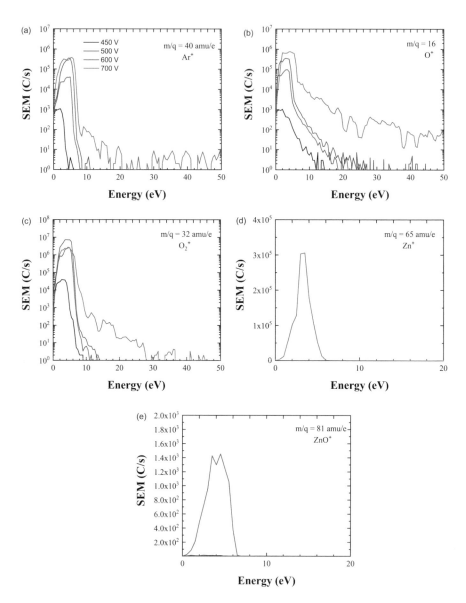

Fig. 6.6 IEDs for (**a**) Ar$^+$, (**b**) O$^+$, (**c**) O$_2^+$, (**d**) Zn$^+$, and (**e**) ZnO$^+$ with different applied target voltages

1.0×10^5 counts/second at 3 eV. The intensities of both populations increased with higher applied voltages. For O$_2^+$ (Fig. 6.6c), the highest intensity was observed for 600 V, while the lowest occurred at 700 V. At 500 V, there were a smaller population of O$_2^+$ ions with energy of 2 eV and a larger population with 5 eV. These energies correspond to the O$^+$ energy distributions at 500 V. The lower ionization of O$_2^+$

(12.1 eV) compared to that of O^+ (13.62 eV) accounted for the order of magnitude higher counts observed for the O_2^+. At an applied voltage of 600 V, the intensity of the O_2^+ IED was still an order of magnitude higher than that of the O^+. At 700 V, however, the intensity of the O_2^+ IED decreased to 5.0×10^5 counts/second, values equal to those observed for the O^+ IED at that voltage. The reduction in O_2^+ at 700 V could be due to dissociation of the molecular oxygen.

The IEDs of the sputtered species (Zn^+ and ZnO^+) are shown in Fig. 6.6d, e. For both the Zn^+ and ZnO^+, the counts were low until the applied voltage reached 700 V. The IED for Zn^+ showed a small population with average energy 2 eV and a much larger population with 3.5 eV energy. The ZnO^+ IED also showed two average energies, one at 3 eV and another at 5 eV. The intensity of the Zn^+ IED was two orders of magnitude larger than that of the ZnO^+; this is understandable given that sputtering occurs primarily on an atomic level, so there will be significantly more Zn atoms available for ionization than ZnO clusters. The negative oxygen ions reported during sputtering of ZnO [45, 46] and HiPIMS of other oxides [47] were not detected in this study or were below the detection limit of the instrumentation arrangements.

Correlation between the observations made in Fig. 6.5a and the evolution of the target current in Fig.e 6.4a provided some insight into how the ion content in the plasma affects the film's microstructural evolution. Films grown with 350 V had poor crystallinity (evident by the broad low-intensity ZnO (002) diffraction peak) due to the low (<1 mA) target current and negligible flux of energetic ionized species to the substrate. At higher applied voltages, the current reached values between 5 A and 17 A, and the substrate was bombarded with Ar^+, O^+, and O_2^+ with energies of 2–6 eV, resulting in films with strong crystallinity and preferred (002) orientation (evident by the narrow high-intensity XRD peak). The films grown at voltages less than 600 V had a slight shift toward lower diffraction angles, which suggested compressive stress in the films. Bombardment by the higher-energy O^+ and O_2^+ that form at voltages of 600 V and larger caused a relaxation of the compressive stress in the films, and this shift in diffraction angle is absent in the film grown at 600 V.

6.4   Conclusions

This chapter discussed part of a larger study to demonstrate low-temperature scalable deposition of ZnO films for TFT applications using reactive HiPIMS. Additionally, this study sought to establish film growth mechanisms, based on correlation between plasma parameters and resultant film structure, to determine the processing conditions needed to synthesize ZnO films with optimal structures for semiconductor devices as determined from review of current literature. Analysis of literature on ZnO TFT device performance showed the need for crystalline structures with the (002) orientation aligned parallel to the substrate plane (i.e., the TFT gate in a back-gated device). To increase the on/off ratio, current density, and saturation current of ZnO TFT channels, smooth films with dense

(002) oriented columnar structures are needed. A review of literature identified that the increased energy of the depositing material and low-energy ion bombardment during HiPIMS could potentially overcome these challenges, but fundamental investigations of plasma properties and their correlation with film structure were absent and critically needed to enhance the current state of the art for ZnO-based devices via a scalable growth route, such as magnetron sputtering.

HiPIMS plasma parameters were studied for sputtering of a Zn target in Ar-O by mapping the ion species and their energies across voltage and pulse processing variables. Sputtering produced low-energy (2–6 eV) ionized gas species (Ar^+, O^+, O_2^+) with currents of $<10^5$ C/s at the substrate surface and 10^3 C/s of Zn^+ at voltages higher than 600 V. The fluxes and relative abundances of ionized species identified in the plasma studies were then correlated with film structure, determined from XRD, AFM, TEM, and SEM, and used to establish mechanisms for film growth and microstructural evolution.

References

1. Ü. Özgür, Y.I. Alivov, C. Liu, et al., A comprehensive review of ZnO materials and devices. J. Appl. Phys. **98**(4), 11 (2005)
2. U. Ozgur, D. Hofstetter, H. Morkoc, ZnO devices and applications: A review of current status and future prospects. Proc. IEEE **98**(7), 1255–1268 (2010)
3. C. Klingshirn, ZnO: Material, physics and applications. ChemPhysChem **8**(6), 782–803 (2007)
4. S. Jeong, B. Park, S. Lee, et al., Metal-doped ZnO thin films: Synthesis and characterizations. Surf. Coat. Technol. **201**(9–11), 5318–5322 (2007)
5. S. Lee, D. Cheon, W. Kim, et al., Combined effect of the target composition and deposition temperature on the properties of ZnO: Ga transparent conductive oxide films in pulsed dc magnetron sputtering. Semicond. Sci. Technol. **26**(11), 115007 (2011)
6. D.C. Look, K. Leedy, L. Vines, et al., Self-compensation in semiconductors: The Zn vacancy in Ga-doped ZnO. Phys. Rev. B **84**(11), 115202 (2011)
7. B. Bayraktaroglu, K. Leedy, R. Neidhard, Microwave ZnO thin-film transistors. IEEE Electron Device Lett **29**(9), 1024–1026 (2008)
8. B. Bayraktaroglu, K. Leedy, Ordered nanocrystalline ZnO films for high speed and transparent thin film transistors, in *Anonymous 2011 11th IEEE International Conference on Nanotechnology. IEEE*, (2011), p. 1450
9. E. Fortunato, P. Barquinha, R. Martins, Oxide semiconductor thin-film transistors: A review of recent advances. Adv. Mater. **24**(22), 2945–2986 (2012)
10. K. Nomura, A. Takagi, T. Kamiya, et al., Amorphous oxide semiconductors for high-performance flexible thin-film transistors. Jpn. J. Appl. Phys. **45**(5S), 4303 (2006)
11. D.C. Look, D. Reynolds, J.W. Hemsky, et al., Production and annealing of electron irradiation damage in ZnO. Appl. Phys. Lett. **75**(6), 811–813 (1999)
12. D.C. Look, J.W. Hemsky, J. Sizelove, Residual native shallow donor in ZnO. Phys. Rev. Lett. **82**(12), 2552 (1999)
13. S. Lany, J. Osorio-Guillén, A. Zunger, Origins of the doping asymmetry in oxides: Hole doping in NiO versus electron doping in ZnO. Phys. Rev. B **75**(24), 241203 (2007)
14. F. Blom, F. Van de Pol, G. Bauhuis, et al., RF planar magnetron sputtered ZnO films II: Electrical properties. Thin Solid Films **204**(2), 365–376 (1991)

15. A. Singh, R. Mehra, N. Buthrath, et al., Highly conductive and transparent aluminum-doped zinc oxide thin films prepared by pulsed laser deposition in oxygen ambient. J. Appl. Phys. **90**(11), 5661–5665 (2001)

16. C. Yu, C. Sung, S. Chen, et al., Relationship between the photoluminescence and conductivity of undoped ZnO thin films grown with various oxygen pressures. Appl. Surf. Sci. **256**(3), 792–796 (2009)

17. S. Amirhaghi, V. Craciun, D. Craciun, et al., Low temperature growth of highly transparent c-axis oriented ZnO thin films by pulsed laser deposition. Microelectron. Eng. **25**(2–4), 321–326 (1994)

18. V. Craciun, J. Elders, J.G. Gardeniers, et al., Characteristics of high quality ZnO thin films deposited by pulsed laser deposition. Appl. Phys. Lett. **65**(23), 2963–2965 (1994)

19. V. Craciun, J. Elders, J.G. Gardeniers, et al., Growth of ZnO thin films on GaAs by pulsed laser deposition. Thin Solid Films **259**(1), 1–4 (1995)

20. L. Han, F. Mei, C. Liu, et al., Comparison of ZnO thin films grown by pulsed laser deposition on sapphire and Si substrates. Physica E **40**(3), 699–704 (2008)

21. S.L. King, J.G. Gardeniers, I.W. Boyd, Pulsed-laser deposited ZnO for device applications. Appl. Surf. Sci. **96**, 811–818 (1996)

22. M. Shimizu, T. Shiosaki, A. Kawabata, Growth of c-axis oriented ZnO thin films with high deposition rate on silicon by CVD method. J. Cryst. Growth **57**(1), 94–100 (1982)

23. K. Haga, M. Kamidaira, Y. Kashiwaba, et al., ZnO thin films prepared by remote plasma-enhanced CVD method. J. Cryst. Growth **214**, 77–80 (2000)

24. W. Maeng, S. Kim, J. Park, et al., Low temperature atomic layer deposited Al-doped ZnO thin films and associated semiconducting properties. J. Vacuum Sci. Technol. B, Nanotechnology and Microelectronics: Materials, Processing, Measurement, and Phenomena **30**(3), 031210 (2012)

25. J. Wang, P. Yang, T. Hsieh, et al., The effects of oxygen annealing on the electrical characteristics of hydrothermally grown zinc oxide thin-film transistors. Solid State Electron. **77**, 72–76 (2012)

26. S. Elzwawi, H.S. Kim, R. Heinhold, et al., Device quality ZnO grown using a filtered cathodic vacuum arc. Phys. B Condens. Matter **407**(15), 2903–2906 (2012)

27. R.J. Mendelsberg, S.H. Lim, D.J. Milliron, et al., High rate deposition of high quality zno: al by filtered cathodic arc. MRS Online Proceed. Libr. Arch. **1315** (2011)

28. X. Xu, S. Lau, B. Tay, Structural and optical properties of ZnO thin films produced by filtered cathodic vacuum arc. Thin Solid Films **398**, 244–249 (2001)

29. F. Van de Pol, F. Blom, T.J. Popma, Rf planar magnetron sputtered ZnO films I: Structural properties. Thin Solid Films **204**(2), 349–364 (1991)

30. R. Menon, K. Sreenivas, V. Gupta, Influence of stress on the structural and dielectric properties of rf magnetron sputtered zinc oxide thin film. J. Appl. Phys. **103**(9), 094903 (2008)

31. K. Kim, J. Song, H. Jung, et al., Photoluminescence and heteroepitaxy of ZnO on sapphire substrate (0001) grown by rf magnetron sputtering. J. Vac. Sci. Technol. A **18**(6), 2864–2868 (2000)

32. S. Jeong, B. Kim, B. Lee, Photoluminescence dependence of ZnO films grown on Si (100) by radio-frequency magnetron sputtering on the growth ambient. Appl. Phys. Lett. **82**(16), 2625–2627 (2003)

33. S. Lee, D. Cheon, W. Kim, et al., Ga-doped ZnO films deposited with varying sputtering powers and substrate temperatures by pulsed DC magnetron sputtering and their property improvement potentials. Appl. Surf. Sci. **258**(17), 6537–6544 (2012)

34. S. Konstantinidis, A. Hemberg, J. Dauchot, et al., Deposition of zinc oxide layers by high-power impulse magnetron sputtering. J. Vacuum Sci. Technol. B: Microelectronics and Nanometer Structures Processing, Measurement, and Phenomena **25**(3), L19–L21 (2007)

35. J. Partridge, E. Mayes, N. McDougall, et al., Characterization and device applications of ZnO films deposited by high power impulse magnetron sputtering (HiPIMS). J. Phys. D **46**(16), 165105 (2013)

36. A.N. Reed, P.J. Shamberger, J. Hu, et al., Microstructure of ZnO thin films deposited by high power impulse magnetron sputtering. Thin Solid Films **579**, 30–37 (2015)
37. K. Matsubara, P. Fons, K. Iwata, et al., Room-temperature deposition of Al-doped ZnO films by oxygen radical-assisted pulsed laser deposition. Thin Solid Films **422**(1–2), 176–179 (2002)
38. U. Helmersson, M. Lattemann, J. Alami, et al., Society of Vacuum Coaters 48th Annual Technical Conference Proceeding. High power impulse magnetron sputtering discharges and thin film growth: A brief review 458–464, (2005)
39. K. Sarakinos, J. Alami, S. Konstantinidis, High power pulsed magnetron sputtering: A review on scientific and engineering state of the art. Surf. Coat. Technol. **204**(11), 1661–1684 (2010)
40. P. Carcia, R. McLean, M. Reilly, et al., Transparent ZnO thin-film transistor fabricated by rf magnetron sputtering. Appl. Phys. Lett. **82**(7), 1117–1119 (2003)
41. A.N. Reed, Reactive High Power Impulse Magnetron Sputtering of Zinc Oxide for Thin Film Transistor Applications. ProQuest in Ann Arbor, MI. The order number is 3710362 (2015)
42. A. Ehiasarian, R. New, W. Münz, et al., Influence of high power densities on the composition of pulsed magnetron plasmas. Vacuum **65**(2), 147–154 (2002)
43. J. Alami, K. Sarakinos, G. Mark, et al., On the deposition rate in a high power pulsed magnetron sputtering discharge. Appl. Phys. Lett. **89**(15), 154104 (2006)
44. A. Hecimovic, A. Ehiasarian, Time evolution of ion energies in HIPIMS of chromium plasma discharge. J. Phys. D **42**(13), 135209 (2009)
45. P. Pokorný, M. Mišina, J. Bulíř, et al., Investigation of the negative ions in Ar/O2 plasma of magnetron sputtering discharge with Al: Zn target by ion mass spectrometry. Plasma Process. Polym. **8**(5), 459–464 (2011)
46. S. Mraz, J.M. Schneider, Influence of the negative oxygen ions on the structure evolution of transition metal oxide thin films. J. Appl. Phys. **100**(2), 023503 (2006)
47. M. Bowes, P. Poolcharuansin, J. Bradley, Negative ion energy distributions in reactive HiPIMS. J. Phys. D **46**(4), 045204 (2012)

Chapter 7
Printed Electronics for Aerospace Applications

Emily M. Heckman, Carrie M. Bartsch, Eric B. Kreit, Roberto S. Aga, and Fahima Ouchen

7.1 Introduction

The printed electronics program at the Air Force Research Laboratory (AFRL) Sensors Directorate explores how the emerging field of additive manufacturing (AM) can benefit electronic devices and applications. The subset of AM we refer to as printed electronics uses technologies such as inkjet printing, aerosol jet printing, and extrusion or micro-dispense printing to selectively deposit conducting and insulating materials to form electronic and photonic devices. The benefits of AM for electronic devices include (1) lower costs, since the materials can be selectively placed and do not need to be coated over the entire substrate and then removed; (2) flexible and lighter-weight devices, since thin and flexible substrates can be used with AM; and (3) rapid prototyping of innovative designs. One of the biggest challenges in printed electronics remains a materials one – designing inks that perform as well as bulk materials. Almost equally challenging is the printing itself: achieving the needed linewidth resolution, surface roughness, and repeatability of results from print to print.

Aerospace applications provide many opportunities for printed electronics to deliver enhanced performance. As unmanned aerial vehicles (UAVs) and attritable platforms continue to dominate Air Force innovations, AM has been able to provide tangible benefits by introducing inexpensive, light-weight, conformable, flexible printed electronic components. Several research areas of the printed electronics program at the Sensors Directorate focus on aerospace applications. These include

E. M. Heckman (✉) · C. M. Bartsch · E. B. Kreit
Sensors Directorate, Air Force Research Laboratory,
Wright-Patterson Air Force Base, OH, USA
e-mail: emily.heckman.1@us.af.mil; carrie.bartsch.1@us.af.mil; eric.kreit.1@us.af.mil

R. S. Aga · F. Ouchen
KBR, Dayton, OH, USA
e-mail: roberto.aga.1.ctr@us.af.mil; fahima.ouchen.1.ctr@us.af.mil

© Springer Nature Switzerland AG 2020   93
M. E. Kinsella (ed.), *Women in Aerospace Materials*, Women in Engineering
and Science, https://doi.org/10.1007/978-3-030-40779-7_7

space-based antennas and antennas for both UAV and attritable platforms. To address the unique challenges of printed components on a space-based antenna, we have sent material samples to the Materials International Space Station Experiment (MISSE) to see how a space environment affects the properties of printed materials.

This chapter will provide an overview of the basic materials challenges facing printed electronic applications – from the conductivity of the materials to the unique post-processing challenges facing printed inks. Because radio frequency (RF) printed electronics rely primarily on conductive inks, we will focus our discussion on these types of materials and not on insulating materials. Next, the characterization techniques used for printed materials will be addressed including the RF characterization needed for antenna applications and the MISSE samples for space-based applications. Finally, we will give an overview of two of the UAV aerospace applications for printed electronics: an all-printed conformal phased array antenna and a data link antenna.

7.2 Materials for Printing

In printed electronics, contacts and interconnects are printed using metal-based particle-containing solutions or suspensions [1]. In the suspension form, metal nanoparticles are used for ink formulations, also called nanoparticle inks. In the solution form, metal organic ions are dissolved in a liquid vehicle (water or organic solvent) leading to a more homogeneous dispersion [1]. There are a variety of metals currently being used in printed electronics. The metals most commonly used in electronic applications are silver (Ag), gold (Au), and copper (Cu). The high cost of Au makes it less attractive despite its known high environmental stability, i.e., resistance to corrosion. Of these metals, Ag is the most commonly used material in printed electronics due to its high electrical conductivity and relative affordability.

Of the various conductive inks available, nanoparticle inks are frequently chosen due to (1) their high particle loadings, which are needed to meet the required percolation thresholds for electrical conductivity; (2) their tunable particle sizing for the desired post-printing process sintering conditions, such as temperature, pressure, and environment; and (3) the ability to choose the ink solvent in either a single or co-solvent system, which allows tailoring of the boiling temperature, vapor pressure, ink viscosity, and substrate surface energy for optimum printing conditions.

Table 7.1 highlights the most common conductive inks printed, tested, and currently used in our lab and their corresponding applications. Conductive polymers, such as PEDOT:PSS, have also been developed for printed electronics applications. These materials offer a modest conductivity at low cost but are limited in terms of chemical and thermal stability. Applications ranging from semi-transparent conductive electrode-to-hole transport layers have been achieved in all printed, flexible photodetectors [2]. Carbon nanomaterials, including carbon nanotubes and graphene, are emerging as alternatives to metals for their high electronic mobility, high

Table 7.1 Printed conductive inks

Material	Manufacturer	Lowest sheet resistance ($\Omega/\square$)	Applications
Silver	Clariant	0.024	Strain gauge, capacitor electrodes, interconnects, CPW lines
Silver	Methode	0.20	Electrodes, interconnects, CPW lines
Gold	UT Dots	0.129	Strain gauge, capacitor electrodes, interconnects, CPW lines
Copper	Lockheed Martin	1.0	Antenna
PEDOT:PSS (polymer)	Sigma-Aldrich	800	Transparent conductive electrode
Carbon nanotube	Brewer Science	11.5	CPW lines, resistors, mid-wave IR absorber
Graphene	Sigma-Aldrich	20.0	Electrodes, interconnects, strain gauge

mechanical flexibility, higher thermal and environmental stability, and potentially lower cost for AM.

7.3   Printing and Post-processing

There are multiple printing techniques available for depositing conductive inks, each corresponding to a specialized printer. The printing techniques that provide the highest resolution, and therefore are best suited to electronics applications, are inkjet printing, aerosol jet printing, and micro-dispense printing. Screen printing, gravure printing, and doctor blade techniques are also common printing techniques that can be used for electronics applications.

In this work, we focus on inkjet, aerosol jet, and micro-dispense printing. Inkjet printing is done by a Fuji Dimatix 16-nozzle inkjet printer. Ink viscosity for inkjet printing must be less than 40 cP, and the narrowest linewidth typically achievable with inkjet printing is 50 μm. Aerosol jet printing is a technique where the ink is aerosolized by placing it in an atomizer and combining it with a nitrogen gas. An Optomec AJ300 is the aerosol jet printer used for this work. The AJ300 can typically achieve minimum linewidths of 15–20 μm, and inks can fall within a much larger viscosity range of 0.7–5000 cP. The micro-dispense printer used is an nScrypt printer. In micro-dispense printing, the ink is pushed out of a syringe by applying a controlled pressure. Inks for micro-dispense printing can have a viscosity anywhere from 1 to 10^6 cP, and typical linewidths tend to average 100 μm, although under certain conditions, linewidths as narrow as 20 μm have been achieved.

After a conductive ink is printed, additional processing is necessary to make the ink conductive. Nanoparticle conductive inks contain solvent additives and surfactants that prevent the nanoparticles from conglomerating and allow the ink to print well, but also prevent the deposited ink from being conductive. These additives must be removed through post-processing. Given the wide range of distinct materials that can be deposited through AM, there is no standard recipe for post-processing. However, the majority of cases can be broken down into a low-temperature drying step, i.e., soft bake, and a high-temperature step, i.e., annealing or sintering. The soft bake step is designed to remove the solvents still present in the material from the deposition process. Temperatures for this step are often less than 100 °C and can be further lowered by drying in a vacuum. For some materials, drying is enough to achieve their intended performance, but many require the additional high-temperature step. For nanoparticle-based inks, this is called sintering. Sintering brings the nanoparticles up to a temperature high enough to allow partial, localized adhesion (called necking) to occur between neighboring particles without having to fully liquefy the nanoparticles. Sintering decreases the void space in the nanoparticle lattice and makes sure that neighboring particles are in good contact with each other. For some inks, such as sol-gel inks, the high-temperature process is not designed to sinter but to increase the crystallinity of the material. The temperature necessary for the high-temperature step varies greatly, typically ranging from 150 to 650 °C, but can be even higher for certain ceramic materials. For most substrate materials (especially polymers), these temperatures would be damaging to the substrates, and thus a simple hotplate or oven bake cannot be used. In these situations, there are three techniques that can be utilized: laser sintering, photonic sintering, and joule heating.

Laser sintering utilizes a focused laser to locally heat the surface of the deposited material without heating the entire substrate. For this technique to work, the deposited material must absorb the wavelength of the laser, and the laser must pass quickly enough to not damage the material. Typically, the focused laser spot size is about 25 microns, and the laser is a single spot that must be pathed serially. This means that for devices with high surface area, the sintering process can be very slow. However, this technique typically achieves a high-quality sinter with results very similar to that of a hotplate bake.

Photonic sintering uses a high-intensity light burst for a very short duration. The most common technique uses a xenon (Xe) arc lamp that delivers an even intensity over a large spectrum. The fundamental principle is that the ink will absorb the emitted light at a greater rate than the substrate material. For an ink that is mostly transparent to the spectrum emitted by the Xe bulb, this technique will not work. There are many variables to control with this technique, including distance from the emitting bulb, pulse intensity, pulse duration, number of pulses, and periodicity of pulses. The goal is to balance all of these variables to provide just enough energy to the material to heat it to the desired temperature. If too much energy is absorbed, it can ablate the material off the substrate. If not enough is absorbed, then the material does not reach the necessary temperature to become activated. Every material and substrate combination needs its own recipe, and for some combinations, the target window is very small, making this technique difficult to use.

Joule heating is fairly restrictive but given the right circumstances can be very useful. In order for this technique to work, the printed material must be at least slightly conductive after the drying step. This technique will not work for many materials, including metallic nanoparticle dispersions that are still fully insulating after deposition and drying. If the material does have some conductivity (even if the resistance is high), an alternating current can be applied that will cause localized heating (joule heating) in the material. Due to the low thermal conductivity of most substrates, the heat tends to stay localized within the deposited material, and rapid sintering can occur. This technique yields a high-quality sinter similar to that of a hotplate or oven bake and can achieve it much more quickly. Additionally, since the heat is generated by the conductive material, it does not cause thermal damage to the substrate.

7.4 Material Characterization

Many aerospace applications either operate in the RF spectrum directly (3 KHz to 300 GHz) or have internal components – such as antennas, transistors, and diodes – that operate in the RF spectrum. To measure the RF response of printed materials, we measure the scattering parameters (S-parameters) on printed films. S-parameters provide a convenient way to look at a system, using both magnitude and phase. For a two-port network, the reflection measurements are S_{11} and S_{22}, and the transmission measurements are S_{12} and S_{21}. S_{11} is a measure of the electrical reflection, and S_{21} is a measure of the electrical transmission.

One method of determining the dielectric constant and permeability for an insulator or semiconductor over a broad frequency range involves obtaining the S-parameter measurements and then using appropriate data processing to obtain the dielectric parameters from the measurements. The specific method we use compares the propagation characteristics of a known reference sample with those of the desired sample. This method determines both the dielectric constant and the loss tangent of the thin film. The technique uses a coplanar waveguide transmission (CPW) line with ground-signal-ground electrodes. From the S-parameter measurements of the CPW line, the capacitance of each sample is determined. From the capacitance and impedance values, the dielectric constant and loss tangents are calculated [3].

To determine if a conductive material is acceptable for specific RF applications, printed CPW lines of these conductive materials are compared to the S-parameter measurements of known CPW lines. Figure 7.1 shows S_{11} for one AU and two AG printed inks. The Au ink is from UT Dots and was printed using the aerosol jet printer. The two Ag inks were Clariant Prelect TPS 35 and Xerox XRCC, and they were printed using the inkjet printer and aerosol jet printer, respectively. Usable conductive materials have S_{11} near the center of the Smith chart, as seen in Fig. 7.1, and S_{21} circling near the outer edge. We have characterized S-parameters for aerosol jet-printed Ag, Au, and Cu and inkjet-printed Ag using various sintering techniques, as well as on a variety of substrates, including photo paper, PET, quartz, silicon, polyimide, and glass.

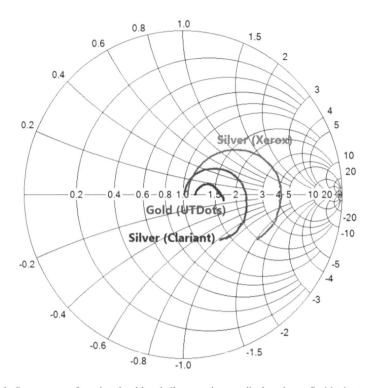

Fig. 7.1 S-parameters for printed gold and silver conductors displayed on a Smith chart

In addition to RF material characterization, there are three basic characterization techniques we use for printed conducting materials that are under consideration for aerospace applications: stylus profilometry for thickness measurements, a four-point probe technique for electrical resistivity measurements, and a combination of stylus profilometry and a high-pulsed current technique for thermal expansion measurements.

Thickness measurements are typically performed after sintering a conductive ink to provide information not only on the material thickness but also on the printer efficiency. These measurements provide an understanding of how the thickness depends on the print parameters and allow users to determine the fabrication time, leading to better ways of building the thick layers needed for some RF and high-power applications. A common method to measure the thickness of printed materials is by stylus profilometry [1], a technique that scans a small stylus with a diamond tip over the printed material – typically a narrow strip printed on a smooth, flat substrate. This measurement technique provides the peak thickness and the average roughness of the printed material.

Electrical resistivity ρ is the most accepted figure of merit for printed conductive materials. The best ρ values for printed traces of metal nanoparticle-based inks are now approximately half that of their bulk counterpart. Measurement of ρ is

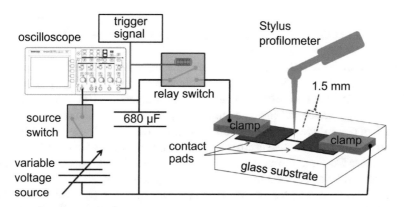

Fig. 7.2 Experimental setup for measurement of current-induced thermal expansion of the printed trace and real-time resistance R(t) of the trace during the application of the pulsed current

traditionally performed by four-point probe technique [4]. In this technique, two of the four probes are used to apply a constant current to the conductive sample, while the remaining probes are used to measure the voltage drop on the sample. When the sample is in a sheet or a thin film, the sheet resistance (R_s) is measured [5]. From the measured sheet resistance, ρ can be calculated by using the equation $R_s = \rho/t$, where t is the measured film thickness.

For high-power current applications, the thermal expansion of a printed trace is important for understanding breakdown at the high-pulsed current regime. To characterize current-induced thermal expansion, we developed an in situ technique that utilizes stylus profilometry [6]. In this technique, shown in Fig. 7.2, the printed trace is subjected to a pulsed, millisecond-range current, while the stylus is positioned at a fixed point on the trace where it continuously monitors height changes. The technique allows the capture of the thermal expansion in real time due to a single current pulse coming from a discharging capacitor. It also enables in situ measurements of the thickness profile before and after applying the pulsed current.

Space-based applications are among the most challenging of aerospace applications to tackle, especially for printed electronics, where many of the materials and inks used are untested in a space environment. MISSE is NASA's flight facility that allows for materials testing in space and is ideal for testing printed materials for spaced-based applications. The MISSE is fixed on the exterior of the International Space Station (ISS) where the mounted samples are exposed to extreme levels of solar radiation, charged-particle radiation, orbital debris, atomic oxygen, hard vacuum, and temperature extremes.

In an effort to gain a baseline understanding of how printed materials react to a space environment, we designed a series of samples that were launched on the MISSE-10 by AlphaSpace. The samples were returned to us in early 2020 after a 6-month exposure to the ISS environment including exposure to extreme heat and cold cycling, ultra-vacuum, atomic oxygen, and high energy radiation. Two identical sample sets were mounted on both the Ram (flight direction) and Wake (opposite

Ram) sides of the MISSE. In addition to the space environment testing, the process of simply submitting the samples showed that the printed materials could withstand the necessary launch conditions.

The aim of the MISSE samples is to test in a space environment the response of various conductive inks from different suppliers. Response is measured in both DC (DC resistivity) and AC (CPW) electrical performance and in a low-pass filter (LCR) using surface mounted, commercially available components to test the strength of conductive epoxies. The inks being tested are Ag (Clariant and Methode) and Au (UT Dots), and the substrates being tested are Rogers CLTE-XT and High Temperature Co-Fired Ceramic from Ferro. The printed samples are then mounted onto thick FR4 slabs for stability using a space-rated epoxy. Figure 7.3a shows the layout of the printed traces on each substrate. One half of each substrate is coated with a 3-4 μm layer of a space-rated passivation layer, CORIN XLS Polyimide from Nexolve. This is a precaution to ensure that some usable data will be retrieved from the samples in case the unshielded materials are completely obliterated by the space environment. At the time of this publication, the MISSE samples have been returned but not yet characterized. Figure 7.3b shows the images of the samples before and after space exposure. A comprehensive study of the effects of the space environment on this sample set is planned for a subsequent publication.

7.5   Applications

RF antenna fabrication is an area that stands to benefit from AM. Traditional antennas are rarely conformal or flexible due to materials and processing constraints, but AM opens up these possibilities to make antennas that are more efficient and lighter weight and that use less space. Additive processes also provide a significant advantage in both freedom of manufacturability and fabrication time, which are both critical to adapting to changing system requirements [1]. For UAV antennas, the advantages offered by AM in terms of reduced fabrication time and lower cost become more attractive because UAVs are often deployed for high-risk missions, and some UAVs are designed not to be recovered at all. Additionally, AM techniques can help achieve multifunctional structures that can simultaneously serve as structural and electrical components; e.g., UAV wings can be designed to have embedded antennas, making the antennas part of the mechanical structure. Traditionally the antennas on airborne platforms are planar, occupy a large footprint, and are installed behind radomes (RF transparent structures designed to protect the antenna apertures). With AM, antennas can be designed and fabricated to be conformally and seamlessly installed into the skins of airborne platforms. This allows for increased installation locations, lighter weight, smaller size, and lower aerodynamic impact on the airborne platform.

One example of a conformal AM antenna can be seen in Fig. 7.4 [7]. This antenna was designed to fit seamlessly into the skin of an existing UAV platform. The substrate material was 3D printed using a polyjet printer and has multiple different radii

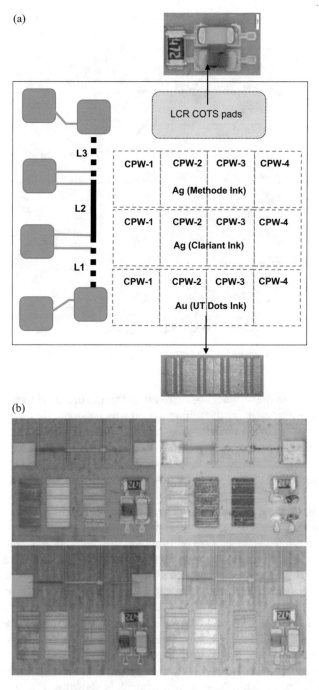

Fig. 7.3 (**a**) Layout of the printed traces for DC and AC material characterization for the MISSE. The three inks tested were Ag ink from Methode (L3), Ag ink from Clariant (L2), and Au ink from UT Dots (L1). The above layout was printed twice for each substrate, with one side passivated and the other left exposed. The photos show how the LCR COTS pads and CPW lines looked after printing. (**b**) Photos of the MISSE samples before launch and after 6 months of space exposure. Top row left is before launch Ram side, top row right is after space exposure Ram side. Bot-tom row left is before launch Wake side, bottom row right is after space exposure Wake side of the MISSE samples before launch and after 6 months of space exposure

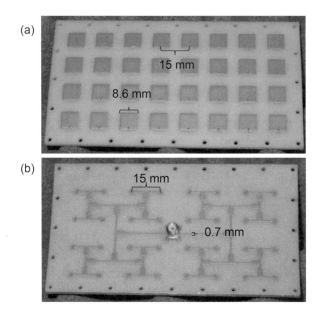

Fig. 7.4 (**a**) Additively manufactured antenna radiating patch elements and (**b**) feed network with SMA connector

of curvature across the sample. The multi-curved nature of the substrate would typically present a challenge for depositing a conductive material (in this example, silver) as there is no simple mathematical description of the surface. To address this, laser mapping was performed to generate a computer definition of the surface that was input to an nScrypt micro-dispense printer. The printer uses the coordinates defined by the laser mapping to ensure that the micro-dispense nozzle remains equidistant from the substrate at all times during the printing. This ensures the highest possible resolution of the resulting print. The radiating patch elements, the feed network, and the vias connecting the two were printed through this method. The results of the printed antenna were promising as they showed very close adherence to simulated planar results, i.e., a maximum error of 7% across the entire tested spectrum. This validates the process of making a conformal antenna utilizing only AM processes and enables new design space for the addition of antennas into complex structures.

A second example of a printed UAV antenna is a fully printed version of an omnidirectional 2.4 GHz data link antenna [8]. Although not conformal or flexible, this antenna could be fabricated using the AM process at a significant cost savings and in less time than its conventionally manufactured counterpart, without sacrificing RF performance. This fully printed antenna is shown in Fig. 7.5. Its design consists of two printed parts glued together with the coaxial cable assembly. For clarity, the two printed parts are labeled as *top hat* and *ground disk*. They are made of ULTEM 1010 printed by a Stratasys Fortus 450mc FDM printer. ULTEM 1010 is a high-strength thermoplastic material that is thermally stable up to 215 °C. It is

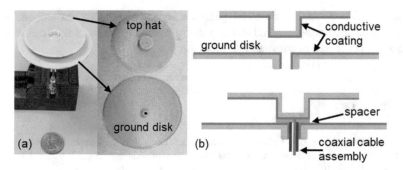

Fig. 7.5 (**a**) Printed data link antenna comprising of two printed parts that are glued together and (**b**) cross section of the fully printed antenna

a good RF material with a dielectric constant of 2.67 and a dissipation factor of 0.001 at 1 GHz. A conductive coating (Dupont silver ink CB028) is printed on one of the sides of the *top hat* and the *ground disk* to serve as the radiating element and ground plane, respectively. Printing was performed using nScrypt micro-dispense printer. After printing, the silver was baked at 120 °C in a convective oven for 30 min. Figure 7.5a shows a photograph of the printed antenna, and Fig. 7.5b shows a cross-sectional diagram describing the details of the antenna. A thermally curable silver epoxy (Creative Materials) was used to connect the center and outer conductors of the coaxial cable assembly to the radiating element and ground plane. It was cured at 120 °C in a convective oven for 60 minutes. Finally, the *top hat* and *ground disk* are attached together using a UV curable adhesive (Norland Optical Adhesive #68). In the attachment process, a 0.5-mm-thick glass spacer was inserted into the gap to maintain a fixed gap value. This is important because the gap spacing between the *top hat* and *ground disk* is crucial to the resonant frequency. The S_{11} and azimuth pattern of the fully printed antenna was measured, and the resonance was found to be at 2.4 GHz, with an S_{11} value of -11.8 dB. The -10 dB bandwidth was 100 MHz. Based on the fabrication time and costs for a conventionally fabricated, similarly functioning antenna, it is estimated that the fully printed version of this antenna costs 10× less and can be made in less than half the time.

7.6   Summary

The AFRL Sensors Directorate has been exploring printed electronics for aerospace applications. Unlike conventional fabrication techniques, where the materials used are well-characterized, a large part of printed electronics research lies in material characterization. We have outlined several characterization techniques used for conductive, printed materials including RF characterization and space environment testing on the MISSE. Our research has identified several UAV applications where printed electronics can make an improvement, and we have discussed two fully printed UAV antennas: a printed RF phased array and an RF data link antenna. Both of these have demonstrated advantages using printed electronics fabrication techniques as compared to conventional fabrication.

References

1. P. Smith, J.G. Korvink, D. Mager, Printed electronics: The challenges involved in printing devices, interconnects, and contacts based on inorganic materials. J. Mater. Chem. **20**, 8446–8453 (2010)
2. R.S. Aga Jr., J.P. Lomardi III, C.M. Bartsch, E.M. Heckman, Performance of a printed photoconductor on a paper substrate. IEEE Photonics Tech. Lett. **26**(3), 305–308 (2014)
3. G. Subramanyam, E. Heckman, J. Grote, F.K. Hopkins, Microwave dielectric properties of DNA based polymers between 10 and 30 GHz. IEEE Microw Wireless Comp. Lett. **15**(4), 232–234 (2005)
4. I. Miccoli, F. Edler, H. Pfnur, C. Tegenkamp, The 100th anniversary of the four-point probe technique: The role of probe geometries in isotropic and anisotropic systems. J. Phys. Condensed Matter **27**, 223201 (2015)
5. F.M. Smits, Measurement of sheet resistivities with the four-point probe. Bell Syst. Tech. J. **37**, 711 (1958)
6. R.S. Aga, E. Kreit, S. Dooley, C. Devlin, C.M. Bartsch, E.M. Heckman, In situ study of current-induced thermal expansion in printed conductors using stylus profilometry. Flexible Printed Electron. **1**, 012001 (2016)
7. A. Piegari, E. Masetti, Thin film thickness measurement: A comparison of various techniques. Thin Solid Films **124**, 249 (1985)
8. K.-L. Wong, F.-R. Hsiao, C.-L. Tang, A low-profile omnidirectional circularly polarized antenna for WLAN access point. IEEE Antennas & Propagation Soc. Symp. **2580**, 1–4 (2004)

Chapter 8
Challenges in Metal Additive Manufacturing for Large-Scale Aerospace Applications

Karen M. Taminger and Christopher S. Domack

8.1 Background: Metal Additive Manufacturing for Aerospace Applications

Over the past decade, additive manufacturing (AM) has emerged as a new process for manufacturing metallic aerospace components [1]. AM processes have matured and become more accepted toward certified aerospace applications. Early adopters have focused on one of two strategies: either a drop-in replacement of a conventionally fabricated component to accumulate service data on secondary structural components, like Boeing's adoption of Rapid Plasma Deposition (RPD™) Ti-6Al-4V galley floor brackets for the 787 [2], or small, intricate parts with geometric complexity such as the fuel injector for GE's LEAP engine [3], turbopumps, and regeneratively cooled rocket nozzles [4]. These early successes have opened up additional designs and opportunities for other components and applications of AM processes across many different materials and scales. As the processes have been approved for use in aerospace applications, interest has also grown for inexpensive, quick production of drop-in replacements of obsolete parts [5].

The applications and service environments for components drive the requirements for selecting the alloys and manufacturing processes. Structures and materials are linked through the manufacturing processes used to produce components. Therefore, an examination of the application and environment contribute to the selection of appropriate combinations of AM processes and materials. Key material and manufacturing characteristics include geometry (size, complexity), mechanical

K. M. Taminger (✉)
NASA Langley Research Center, Hampton, VA, USA
e-mail: karen.m.taminger@nasa.gov

C. S. Domack
Analytical Mechanics Associates, Inc., Hampton, VA, USA

© Springer Nature Switzerland AG 2020   105
M. E. Kinsella (ed.), *Women in Aerospace Materials*, Women in Engineering
and Science, https://doi.org/10.1007/978-3-030-40779-7_8

properties (strength, stiffness, durability, and damage tolerance), density, and resistance to chemical and thermal environments.

In consideration of structural aerospace applications, several common characteristics emerge. Aircraft fuselage and cryogenic tanks for rockets are typically aluminum or aluminum-lithium alloys, driven by the requirements for lightweight, sealed pressure vessels that are damage tolerant at ambient and cryogenic temperatures. Smaller fuel tanks for satellites and space exploration vehicles (like planetary landers and rovers) tend to be fabricated from titanium alloys for higher strength-to-weight ratios and chemical resistance to corrosive or reactive fluids. Low production rates favor the development of custom, conformal tanks to maximize useable volume. Rocket nozzles are exposed to extreme temperature changes, from cryogenic fuels to hot exhaust gases, while also requiring strength and stiffness to sustain the high pressure, high vibration loads over a relatively short duration of a launch. Durability is becoming more important as an increasing number of rocket components are being designed for recovery and reuse after a launch. All of these components carry structural loads in a variety of environmental conditions with long service life expectancies and are large to very large in scale (with dimensions measured in feet or meters).

8.2   Attributes of Additive Manufacturing Processes

Additive manufacturing has been classified into seven general classes based upon attributes of how the feedstock materials are turned into a part in a layer additive manner [6]. Two general classes of AM processes that are most mature and germane to metallic aerospace applications are powder bed fusion (PBF) and directed energy deposition (DED). Each class has its benefits and drawbacks related to feature size, deposition rate, and substrate variations that dictate which process is more suitable for specific applications.

Powder bed fusion processes scan an energy source, typically either a laser or an electron beam, over a thin layer of powder spread across a powder bed. This is a fusion process that creates a small (approximately 0.04 in or 1 mm) diameter molten pool that is scanned quickly (up to 16,500 in/min or 7 m/sec) across the powder bed to fuse a layer of powder to the underlying layers. PBF processes are capable of building small parts with a high degree of complexity because of the size of the molten pool, and the powder acts as a support material. Despite the fast beam scan speeds, the size of the molten pool and powder layers result in a relatively slow deposition rate [6]. PBF processes are ideal for integrating smaller components for aircraft and rocket engine and nozzle applications with a high degree of intricate detail, such as cooling passages.

Directed energy deposition (DED) processes use either blown powder or wire feedstock and one of a variety of energy sources, including electron beam, laser, arc, or plasma to deposit layers onto a substrate. DED processes typically move at significantly slower rates than PBF processes, because either the deposition head or the substrate is being translated, as compared to rapidly scanning an energy beam in the PBF processes. DED processes are capable of higher deposition rates, because they can more readily increase the power and the feedstock feed rates. The trade-off for higher deposition rates is the need to use a larger melt pool, typically 0.1–0.25 inches (4–10 mm) in diameter, which sacrifices fine details to build parts faster. In many cases, the use of higher deposition rates followed by a finishing machining step still trades favorably for manufacturing time and cost, particularly for tooling and aerospace parts which require a finish machining step to achieve a surface finish suitable for fatigue-driven applications.

Another benefit of wire-fed DED processes is the ability to rapidly change alloys between different part production runs, with little potential for contamination because the wire feedstock is self-contained. Changing feedstock simply requires changing the wire spool. With a dual wire feeder arrangement, wire-fed DED is also amenable to feeding two wires of different compositions simultaneously. This enables deposition of functionally graded structures through differential control of individual wire feed rates to adjust the composition during the build. Powder beds and blown powder systems require extensive cleaning of the powder storage and feeding systems to avoid cross-contamination when changing from one alloy to another.

For PBF processes, parts are typically built on a flat baseplate to facilitate spreading an even layer of powder across the bed for the deposition process. Any variations in the height of the substrate will interfere with the powder spreader. DED processes operate in a multi-axis environment where the deposition heat source and feedstock move relative to the part. Therefore, DED processes can be used to add details onto simplified preforms or existing parts to effect repairs. In these instances, the substrate and the deposited material will constitute the final part, which requires paying specific attention to the microstructure and mechanical properties of the interfaces to ensure continuity and consistent structural performance.

8.3   Directed Energy Deposition: Electron Beam Freeform Fabrication (EBF³)

In 2002, researchers at NASA Langley Research Center (LaRC) started developing metal additive manufacturing processes. With a specific interest in low-rate production of large-scale metallic aerospace structures, the focus was on development of a directed energy deposition (DED) process using an electron beam heat source and wire feedstock called electron beam freeform fabrication (EBF³). Electron beams offer excellent energy coupling to metallic substrates and are well-suited for operating in a vacuum. Wire feedstock avoids problems with powder migration in reduced gravity and enables more precise control. EBF³ was selected based on the greatest benefit to NASA's research objectives, including deposition of aluminum, titanium, and nickel-based aerospace alloys and efficient operation in a space environment.

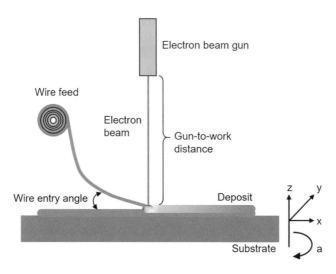

Fig. 8.1 Schematic drawing of key components of the EBF³ process. (Image credit: NASA)

EBF³ uses a focused electron beam in a vacuum environment to create a molten pool on a metallic substrate. The beam is translated with respect to the surface of the substrate, while wire is fed into the molten pool. The deposit solidifies immediately after the electron beam has passed, having sufficient structural strength to support itself. The sequence is repeated in a layer additive manner to produce a near-net-shape part needing only post-deposition heat treatment and finish machining. A schematic of the EBF³ components is shown in Fig. 8.1; note that the design of different systems may change which components are moving, but the fundamentals of the process remain the same.

There are many key EBF³ process variables that may be grouped into three categories. First is the geometry of the hardware, which includes the wire entry angle into the molten pool and the gun-to-work distance (the distance between the electron beam gun and the molten pool). Second is the kinematics of the system, which includes the tilt and rotation of different axes, the translation speed and direction of the molten pool relative to the wire entry, and the wire feed rate. Third is the beam parameters, which includes the accelerating voltage, beam current, focus, deflection, and raster. The number of variables offers a breadth of combinations that can be tailored to optimize the final shape and microstructure of the deposited material [7–10].

The operational concept of EBF³, illustrated in Fig. 8.2, is to build a near-net-shape metallic part directly from a computer-aided design (CAD) file without the need for molds or tooling dies. Current computer-aided machining practices start with a CAD model and use a post-processor to write the machining instructions (G-code) defining the cutting tool paths needed to make the part. EBF³ uses a similar process, starting with a CAD model, numerically slicing it into layers, and then

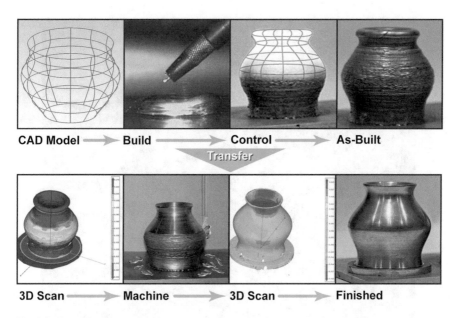

CAD Model ➝ **Build** ➝ **Control** ➝ **As-Built**

Transfer

3D Scan ➝ **Machine** ➝ **3D Scan** ➝ **Finished**

Fig. 8.2 Operational concept of the electron beam freeform fabrication (EBF³) process. (Image credit: NASA)

using a post-processor to write the G-code defining the deposition path and process parameters for the EBF³ equipment.

There are three EBF³ systems at NASA LaRC that are being developed for different purposes and experiments. The first is the original large-scale system based upon a commercial electron beam welder from Sciaky Inc., shown in Fig. 8.3. This system includes a 60 kV/700 mA electron beam gun and dual wire feeders capable of independent, simultaneous operation. Positioning is programmable through six axes of motion (X, Y, Z, gun tilt, and positioner tilt and rotate), with a build envelope of $60 \times 24 \times 24$ inches ($1.5 \times 0.6 \times 0.6$ m). The large-scale EBF³ system is inside a vacuum chamber that operates in the range of 5×10^{-5} torr (6.5×10^{-3} Pa). Due to its size and vacuum pumping capacity, this is the primary system used for process development and for large-scale deposition at NASA LaRC.

The second EBF³ system (Fig. 8.4) is portable and comprises a small vacuum chamber, a 30 kV/100 mA electron beam gun mounted fixed in the top of the chamber, four-axis motion control system on the table (X, Y, Z, and rotation), single wire feeder, and data acquisition and control system. This portable EBF³ system was used to study the effect of microgravity on build geometry and solidification microstructure during parabolic flights on NASA's C-9 [11]. The third EBF³ system (Fig. 8.5) is the second-generation portable system. This is similar in size to the first-generation portable system, but has a movable 20 kV/100 mA electron beam gun inside the vacuum chamber with a fixed table and single wire feeder. This system was intended to test smaller-scale electron beam guns that are contained within

Fig. 8.3 Large-scale EBF³ system at NASA LaRC for process development. (Image credit: NASA)

Fig. 8.4 First-generation portable, fixed-gun EBF³ system undergoing 0-g testing in parabolic flights. (Image credit: NASA)

the vacuum chamber to more accurately replicate conditions that would exist in space.

There are several attributes to the EBF³ process that are attractive for high rate deposition. Electron beams couple efficiently with electrically conductive materials like metals, enabling high beam energies that can be focused and rastered using electromagnetic lenses, as shown in Fig. 8.6. This characteristic enables greater control over the melt pool width and depth and allows for a high volume of wire to be melted during a single deposition pass.

Fig. 8.5 Second-generation portable EBF³ system with movable gun inside the vacuum chamber. (Image credit: NASA)

Fig. 8.6 Schematic drawing of an electron beam gun. (Image credit: NASA)

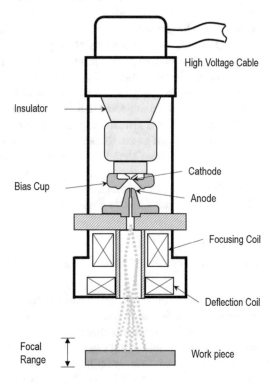

The electron beam requires vacuum to sustain a focused electron beam over long gun-to-work distances. This results in a clean environment, free of the risk of absorbing gases from the air without requiring consumables like argon, as is required for PBF processes. However, since this is a fusion process, selective vaporization of highly volatile alloying additions has been documented as a result of the vapor pressure differences in vacuum [12]. This can result in alloying chemistries that fall outside of the acceptable range, which affect the microstructural evolution and mechanical properties [13]. A simple enrichment of the starting feedstock chemistry to adjust for the anticipated alloying losses is common in weld wire compositions and may be equally applied for many alloys used in EBF³.

8.4   Case Studies: Challenges in EBF³ Deposition

Some of the initial parts built at NASA LaRC using EBF³ were generic rocket nozzle geometries. In their simplest form, rocket nozzles may be built as a thin-walled body of revolution with gradual transition angles to produce the desired nozzle shape. Rocket nozzles are attractive for EBF³-type processes because they are larger parts that can be produced more cost-effectively than with processes requiring non-recurring tooling costs, especially at the prototype and testing phases. The ability to quickly build functional prototypes without tooling enables rapid turnaround of different geometries to validate analytical performance predictions and iteratively refine and optimize the geometries in a reasonable time and cost.

Rocket nozzles are subjected to harsh environmental conditions including intense temperature extremes, structural loads, and exposure to oxidation, corrosion, and erosion from exhaust gases. Rocket nozzles are only exposed to these severe conditions for a short time; therefore strength, stiffness, toughness, and thermal and chemical resistance are key characteristics for surviving the harsh launch environment. These design considerations lead to selection of titanium- or nickel-based alloys for many rocket nozzle applications.

8.4.1   Generic Inconel® 718 Rocket Nozzle

The first attempt at using EBF³ to build a rocket nozzle used Inconel® 718, a high-temperature nickel-based alloy commonly used in aircraft and rocket engine applications. Due to its complex composition and microstructure, this alloy is difficult to weld and is susceptible to hot cracking. The build geometry selected was one bead-width wide, resulting in an as-built wall thickness of 0.25–0.30 inches (6.3–7.6 mm). The nozzle started with a 10-inch (25.4 cm)-diameter cylindrical section that narrows and widens in a generic nozzle geometry, built to a height of 18 inches (45.7 cm). Photos of the Inconel® 718 nozzle are shown in Fig. 8.7.

Fig. 8.7 Generic Inconel® 718 rocket nozzle in large EBF³ system at NASA LaRC (left) and after partial machining (right). Note use of a fixed build platform (no rotary motion) for EBF³ deposition. (Image credit: NASA)

In most cases, the part to be built using EBF³ is located below the electron beam gun and built in a vertical layering direction, as shown in Fig. 8.7. Height in the large EBF³ system is limited with the tilt-rotary table installed; therefore the table was removed to provide greater build Z-height. This required the part to be built using all X-Y moves rather than a rotary move with adjustments in either X or Y to obtain the variable geometry. Wire entry into molten pool is oriented such that the wire is aligned with the X-axis and feeds directly into leading edge of molten pool on +X moves. This means as the deposition takes place and the electron beam gun translates around the build path, the wire orientation feeds into the side of molten pool in +Y and −Y moves and into the trailing edge of molten pool on −X moves.

In the current EBF³ configuration, wire is fed into the molten pool at a relatively low entry angle, 30–35° from the baseplate. This wire entry angle is dependent upon the hardware geometry of the wire feeder and the gun-to-work distance but is aligned so that the wire enters the molten pool on the substrate at the same location as the desired focal point of the electron beam. The wire is heated as it passes through the electron beam into the molten pool. When feeding into the leading edge (+X move), the wire has sufficient time in the molten pool that it fully melts during the transit time. When feeding into the side (±Y moves), the molten pool tends to widen slightly. Since the wire feed rate dictates the volume of wire feeding into the molten pool, for a constant wire feed rate, a slight widening of the molten pool also means that the layer height is proportionally lower. For a single layer, this is

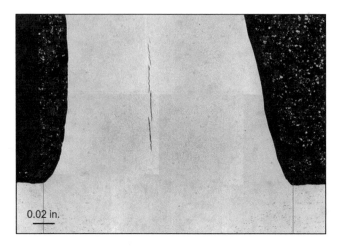

Fig. 8.8 Cross section of an Inconel® 718 EBF³ deposit with evidence of hot cracking at the centerline. (Image credit: NASA)

imperceptible, but the layer height error will build up over time if uncorrected. This is evident in the Inconel® 718 nozzle shown in Fig. 8.7: an uneven build was noted at the shoulder of the narrowest part of the nozzle, which resulted in a slight kink in the axis of the nozzle and rendered this particular component unusable. To correct for this phenomenon, change of the wire entry angle and beam deflection may be used to mitigate the height variations in parts that cannot be built using the rotary table.

Since EBF³ occurs in a vacuum environment, cooling occurs primarily by conduction through the build plate into the support table. A little heat is also expelled through radiation, but no convective cooling is possible due to the absence of an atmosphere. This results in lower cooling rates than other wire-fed AM processes that are performed in argon or air (such as laser-based, plasma, or arc-based processes), but EBF³ cooling rates can still result in hot cracking [14].

During fabrication of the Inconel® 718 nozzle, cooling rates were high enough to induce hot cracking. The basic mechanism of hot cracking results when the nickel freezes quickly into dendrites, expelling the alloying elements like niobium and molybdenum into the interdendritic regions. The niobium-rich interdendritic region is under high thermal residual stresses and is a lower-strength phase, leading to preferential cracking in that region. Cracks in the centerline of the single-bead deposit were observed during initial deposition parameter development for the Inconel® 718, shown in Fig. 8.8.

Weld bead shape dictates the weld metal solidification pattern and, in turn, is influenced largely by welding parameters. To address the hot cracking in the Inconel® 718, the beam power and travel speed were both increased, while the part was programmed for continuous building in a helical pattern (i.e., no starts and stops as each X-Y layer is completed). The process was driven such that 1–2 inches (2.5–5 cm) of build height was glowing cherry red during deposition. This resulted

Fig. 8.9 Generic Ti-6Al-4V rocket nozzle as-built (left) and after machining (right). (Image credit: NASA)

in a heat soak that allowed enough time for the niobium and molybdenum to diffuse back into solution, eliminating the low-strength interdendritic region and thereby the hot cracking.

8.4.2 Generic Ti-6Al-4V Rocket Nozzle

The next attempt at using EBF³ to build a rocket nozzle used Ti-6Al-4V, which has excellent corrosion resistance and is suitable for use temperatures up to around 775 °F (415 °C). The primary benefit of using Ti-6Al-4V is the significantly lower density than nickel-based alloys. The rocket nozzle geometry selected was also one bead-width wide, resulting in an as-built wall thickness of 0.20–0.25 inches (5–6.3 mm). This nozzle was built starting with a 4-inch (10 cm)-diameter base, converging and then diverging to 8 inches (20 cm) at the top, with a final height of nominally 14 inches (36 cm). As was the case with the Inconel® 718 nozzle, the Ti-6Al-4V nozzles were too tall to use the tilt-rotary table; hence they were built on a lower-profile table with all X-Y programmed moves. Photos of the Ti-6Al-4V nozzle are shown in Fig. 8.9.

The geometry of the Ti-6Al-4V nozzle had a more aggressive unsupported overhang than the Inconel® 718 nozzle. With the tilt table, the geometry could have been built, so the table was tilted to maintain orthogonality of the electron beam and deposited bead to the contour of the nozzle. Without the tilt table, the deposited bead had to be stepped over from the previous bead to achieve an unsupported overhang. This means that the viscosity and surface tension of the molten pool dictate the maximum unsupported overhang that can be achieved. For the Ti-6Al-4 V, the viscosity and surface tension are low enough that a 45° angle is achievable just by

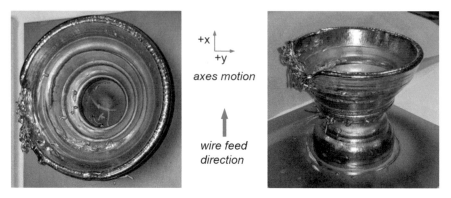

Fig. 8.10 Initial attempt at Ti-6Al-4V nozzle experienced issues with wire entry angle, showing a top view (left) and side view (right). (Image credit: NASA)

stepping the layers over by up to 60% of the bead width on each subsequent pass. The molten pool perches on the edge of the previously deposited layer to achieve the overhang. Care must be taken not to run the process too hot in these cases, because that will increase the fluidity, melt too deep into prior layers, and jeopardize successful adherence of the molten pool to the part. Therefore, the EBF³ process is operated at slightly lower power levels for these geometries to enable maximum unsupported overhang angles.

The Ti-6Al-4V nozzles were built with *X-Y* programmed moves in a counter-clockwise direction. Variations in bead width and height were observed as with the Inconel® 718 nozzle, due to the changing wire entry angle into the molten pool. The typical Ti-6Al-4V EBF³ deposition parameters were modified to enable building the unsupported overhang, but this created other problems during the first trial for this nozzle. Wire feeding into the trailing edge of the molten pool (−*X*), particularly with the lower layer height, results in the wire not always having sufficient time to melt. The wire can deflect off the bottom and out the side of the molten pool when the process is running at lower power levels, as can be seen in Fig. 8.10. Note that this only occurred in the –*X*, –*Y* quadrant of the build, recovering as the programmed moves trended back toward the +*Y* direction. To correct for this error, rotary moves with the wire feed angle held constant are the preferred solution. In instances when a rotary move is not possible, a slight increase in beam power provides the additional energy necessary to fully melt the wire, even in the –*X*, –*Y* quadrant, enabling successful fabrication of the component.

EBF³ deposition of Inconel® 718 is different from Ti-6Al-4V. Unsupported overhangs are much harder to achieve with Inconel® alloys, as compared to Ti-6Al-4V. Even with stepping over the programmed layer position in an *X-Y* move, the molten pool tends to pull back to the position of the prior deposited layer. The end result is that Inconel® 718 (and other Inconel® alloys like 625) are only able to sustain lower unsupported overhangs when built in a purely *X-Y* move. To achieve higher unsupported overhangs, use of the tilt table is necessary to tilt the part and physically change the edge of the molten pool onto which the electron beam is directed.

8.4.3   Copper-Nickel Bimetallic Nozzle

A complete liquid-fueled, regeneratively cooled rocket motor specifically designed for AM was built and successfully test-fired by NASA in 2017. This particular design incorporated internal cooling passages that conform to the nozzle throat geometry to actively cool the nozzle, with different portions being built using copper for thermal conductivity and nickel for strength. The design leveraged the fine feature detail of laser powder bed fusion (LPBF) and the high deposition rate, multi-material capability of EBF[3] to produce a near-net-shape, bimetallic structure that could not be built solely by either process [4].

The overall geometry of this chamber and nozzle assembly was approximately 9 inches (23 cm) in diameter and 18 inches (46 cm) in length. A GRCOP-84 liner with internal cooling channels was fabricated via the LPBF process at NASA's Marshall Space Flight Center in two sections; the overall length of the component precluded building the liner as a single piece. The sections were welded together, and then EBF[3] was used to add the Inconel® 625 structural jacket. GRCOP-84 is a high-temperature copper-based alloy developed at NASA Glenn Research Center for combustion chamber liners of regeneratively cooled rocket engines, with approximately 6.6 wt% Cr and 5.7 wt% Nb that forms Cr_2Nb precipitates for increased strength [15]. Inconel® 625 is a weldable, high-strength, high-temperature nickel-based alloy commonly used in aircraft and rocket engines.

The EBF[3] deposition portion of this effort required development of several new approaches. Internal, collapsible stainless steel tooling was developed to support the inside of the GRCOP-84 liner in the EBF[3] system, as shown in Fig. 8.11. The nozzle was fixtured using the rotary table with a 90° tilt to operate similar to a lathe. Since

Fig. 8.11 Copper-nickel bimetallic nozzle in the large EBF[3] system at NASA LaRC (left) and after completion of EBF[3] deposition (right). (Image credit: NASA)

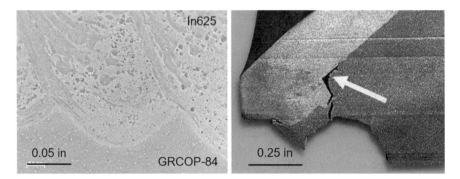

Fig. 8.12 Cross-section micrograph showing the Inconel® 625 bead interface with the GRCOP-84 liner (left) and section of the aft end of the nozzle showing the limitation of the wire feed angle on the EBF³ system preventing access to the 90° corner feature on the GRCOP-84 liner (right). (Image credit: NASA)

the EBF³ deposition was directly onto the GRCOP-84 liner, this configuration was able to maintain the nearly orthogonal orientation of the electron beam and wire relative to the surface of the part. A tailstock assembly was used to support the outer end of the internal tooling, since the entire assembly weighed in excess of 750 pounds (340 kg) (including the tooling, nozzle liner, and structural jacket). The X- and Z-axes were programmed to follow the contour of the nozzle liner, while the rotary table continuously rotated the part during deposition. This configuration allowed for the wire to feed into the side of the molten pool to obtain a consistent bead width and height throughout the part fabrication. Additionally, the part was programmed for helical builds to reduce the number of starts and stops.

The deposition of the 0.3-inch (7.5 mm)-thick Inconel® 625 structural jacket was relatively straightforward and was based on a set of optimized deposition parameters that had been developed during preliminary experiments. The initial deposit layer required significantly higher beam power to bond the Inconel® 625 to the GRCOP-84, due to its higher thermal conductivity. After the initial layer of Inconel® 625 was deposited onto the GRCOP-84, all subsequent layers used lower beam power to deposit on top of the Inconel® 625 of the previous layers. Figure 8.12 shows a cross section of the interface between the GRCOP-84 liner and the Inconel® 625 jacket. The scalloped features are the side-by-side beads of Inconel® 625 where they melt into the GRCOP-84. Deposition of the first layer of Inconel® 625 onto the GRCOP-84 required melting of the substrate to obtain a metallurgical bond. Traces of copper can be seen mixing in the initial layer of Inconel® 625. Chemical analysis has shown this layer to be as much as 30% copper, which allows for a slight transition zone between the two alloys. This was desired to assist with grading the materials, since the two alloys have drastically different thermal conductivities, coefficients of thermal expansion, and elevated temperature strengths. In the circumferential direction of the rocket nozzle, this interface worked well.

The geometry of the LPBF GRCOP-84 at the ends of the part proved to be challenging, due to the entry angle of the wire feed into the molten pool in the EBF³

Fig. 8.13 End effects produced high residual stresses that resulted in separation of the Inconel® 625 during deposition onto a GRCOP-84 extrusion. (Image credit: NASA)

process. Some of the sharper corners were inaccessible, leaving areas where the Inconel® 625 was not bonded to the GRCOP-84. These regions were identified, and the LPBF geometry was subsequently modified to make more gradual angles that were accessible to the EBF³ wire feeder. Figure 8.12 shows a cross section of the regions in question, which were excised from the part for metallurgical analysis after completion of the deposition.

Coupons for interface tensile strength testing were desired to characterize the interfacial region in this bimetallic rocket nozzle. An initial attempt involved EBF³ deposition of Inconel® 625 onto an extruded GRCOP-84 block measuring approximately 8 × 2 × 2.5 inches (20 × 5 × 6.3 cm). The intent was to deposit a brick approximately 7 × 1.5 × 2.5 inches (18 × 3.8 × 6.3 cm) of Inconel® 625 onto the 2-inch (5 cm)-wide face of the GRCOP-84 extrusion. During deposition, thermal residual stresses from the end effects and contraction of the Inconel® 625 caused the Inconel® 625 deposit to separate from the GRCOP-84 block at the interface after only seven layers (~0.7 inches (1.8 cm)) had been applied. Since the cooling path was interrupted where the separation occurred, the end of the Inconel® 625 that peeled up is glowing cherry red in Fig. 8.13. This experiment demonstrated that high residual stresses develop during the EBF³ process as a result of the differences in thermal conductivity, thermal expansion, and strength at elevated temperatures between the GRCOP-84 and the Inconel® 625.

It was recognized that the stress state of this interface block is not representative of the stress state in a cylindrical deposition. Therefore, the GRCOP-84 extrusion from Fig. 8.13 was machined into a 2-inch (5 cm)-diameter, 8-inch (20 cm)-long solid cylinder. A 2-inch (5 cm)-thick layer of Inconel® 625 was deposited onto the outside of the solid GRCOP-84 cylinder, as shown in Fig. 8.14. The EBF³ layers

Fig. 8.14 Successful deposition of GRCOP-84-Inconel® 625 interface coupons for tensile testing. (Image credit: NASA)

were deposited in overlapping rings, with a start and a stop on each bead, stopping periodically to allow the part and EBF³ gun to cool. Slices were cut from the cylinder, and flat dog-bone specimens were machined with the GRCOP-84-Inconel® 625 interface at the center of the dog-bone specimens. Tensile tests revealed that failures generally occurred in the copper substrate, except where a localized defect occurred at the interface between the first and second layers of Inconel® 625. More detailed evaluation of the effects of copper uptake in the strength and the residual stress state at this complex interface is underway.

8.4.4 Aluminum Stiffened Panels for Aerostructures and Launch Vehicles

Recent structural design optimization studies have focused on improving the performance of aircraft fuselage and wing structures. One approach has been to use curvilinearly stiffened skin panels to tailor the properties of the structure to simultaneously reduce weight and provide improved stiffness or acoustic damping. Thin, lightweight aircraft wings are susceptible to aeroelastic instability and flutter. Structural tailoring to change the bend-twist stiffness behavior of the wing may be accomplished with curvilinear spars, ribs, and stiffeners [16]. Other optimization studies have explored use of curvilinear stiffeners to passively induce destructive interference between acoustic waves to reduce cabin noise or acoustic fatigue loads in launch vehicles and supersonic missile housings [17].

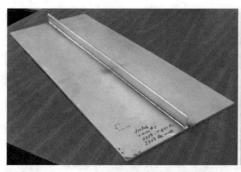
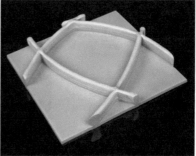

Fig. 8.15 Examples of 2219 Al straight (left) and curvilinear (right) stiffeners on skins representative of aircraft and launch vehicle stiffened skin panels; the panel on the left is approximately 8 × 18 inches (20.3 × 45.7 cm) with a 1-inch (2.5 cm)-tall stiffener. (Image credit: NASA)

The advent of large-scale DED processes may enable manufacturing of new structural designs such as these studies have proposed. The addition of features onto simple forgings, castings, or rolled product forms can reduce manufacturing time and costs and enable improved performance. The size of parts that can be produced via EBF³ is limited only by the size of the vacuum chamber. To study the fundamental issues with combining conventional and AM approaches, EBF³ has been used to deposit 2219 Al blade stiffeners onto 2219 baseplate. The baseplate for the EBF³ deposition process will become a structural skin with deposited integral stiffeners. Straight and curvilinearly stiffened panels, as shown in Fig. 8.15, have been built to evaluate the effects of the repeated traverses of a localized heat source during layer additive processes on the resulting microstructures, mechanical properties, structural distortion, and residual stresses.

In fusion DED processes like EBF³, the first layer melts into the baseplate, introducing a heat-affected zone in the area immediately surrounding the interface between the deposited bead and the baseplate. Each subsequent layer melts into the previous layer or two to provide good adhesion. The heat input spikes as the beam passes by and then dissipates into the surrounding structure. As layers continue to build up, the relative temperature spikes are reduced in magnitude, but the background temperature continues to rise. The magnitude and rate of the temperature rise depends on the size of the part and the time between passes. This complex thermal profile drives the microstructural evolution during the EBF³ deposition process. In the micrograph shown in Fig. 8.16, the heat-affected zone and variations within the deposited layers are clearly evident in the cross section of a 2219 Al stiffener on 2219 Al 0.25-inch (6.3 mm)-thick baseplate.

Another challenge that must be monitored and mitigated with a localized moving heat source is the development of thermal residual stresses that can result in warping or distortion of the deposited part or the substrate on which it is being deposited. This is a particularly important issue when the substrate is a simplified preform onto which details are deposited, such as bosses, flanges, or stiffeners. Various distortion mitigation techniques have been studied, including preheating or actively cooling

Fig. 8.16 Cross section of an as-deposited 2219 Al EBF³ deposit on 2219 Al sheet clearly shows differences in microstructures and heat-affected zones resulting from the EBF³ process. (Image credit: NASA)

0.05 in.

the substrate, thicker weld land regions, and pre-stressing the baseplate in the opposite direction to compensate for the anticipated distortion [18].

Although distortion can be mitigated through different deposition and fixturing approaches, the issue of dissimilar microstructures and heat-affected zones presents additional concerns. The base structures may be heat treated to a high strength condition, such as a –T6 or –T8 temper in aluminum alloys, but the deposited material and surrounding heat-affected zone are closer to an annealed (-O temper) condition in the as-deposited condition. AM of aluminum alloys is challenging because they are not used in an annealed condition. High-strength aerospace aluminum alloys are precipitation strengthened and derive their strength from a combination of heat treatment and work (either from a forming or rolling operation or from a stretch imparted after heat treatment) to introduce dislocations. AM-deposited aluminum alloys tend to have strengths between the annealed and solutionized and naturally aged (–T4 temper) properties as a result of the microstructures that develop during the thermal excursions of the AM processes [19]. Studies have shown properties near –T6 are achievable through a direct age after deposition in 2219 and 2139 alloys. This avoids the need for a high-temperature solutionization and quench to minimize distortion, but fails to introduce the dislocation strengthening that is typically introduced through a stretch to obtain –T8 properties [20].

Alloy development to exploit the processing conditions, such as thermal cycling during the EBF³ deposition process, is an active area of research for future aerospace aluminum alloys. Once the thermal differential and distortion issues can be solved, selectively integrating large-scale additive manufacturing into components fabricated with conventional processes offers affordable, manufacturable solutions applicable to commercial production.

8.5 Concluding Remarks

Manufacturing is the link between structures and materials that can be used to form an integrated solution enabling advances in performance, cost, or manufacturing time. Metal additive manufacturing approaches like LPBF and EBF[3] are two of many tools in the manufacturing toolbox that add to the solution design space. Presently, metal AM processes are well-suited for prototypes and research, and lessons learned in one process space are often applicable for other AM processes. High deposition rate of DED processes in combination with other advanced manufacturing processes offers a new approach to fabricate unconventional large-scale aerospace structures. The case studies presented in this paper were useful for identifying challenges and developing robust processing parameters and techniques to mature the EBF[3] process for wider application toward qualification for adoption into production.

Acknowledgments The authors wish to acknowledge the contributions of Robert Hafley and Richard Martin at NASA Langley Research Center for their support in developing the EBF[3] process and associated research presented in this document.

References

1. W.J. Sames, F.A. List, S. Pannala, R.R. Dehoff, S.S. Babu, The metallurgy and processing science of metal additive manufacturing. Int. Mater. Rev. **61**(5), 315–360 (2016)
2. H. Canaday, Making 3D-printed Parts for Boeing 787s. Aerospace America, September 2018 (2018)
3. T. Kellner, The FAA Cleared the First 3D Printed Part to Fly in a Commercial Jet Engine From GE. GE Reports. http://www.gereports.com/post/116402870270/the-faa-cleared-the-first-3d-printed-part-to-fly/ (2015)
4. P. Gradl, S.E. Greene, C. Protz, B. Bullard, J. Buzzell, C. Garcia, J. Wood, Cooper, K., Additive Manufacturing of Liquid Rocket Engine Combustion Devices: A Summary of Process Developments and Hot-Fire Testing Results. Proceedings of the 54th AIAA/SAE/ASEE Joint Propulsion Conference, Cincinnati, OH, (paper no. AIAA-2018-4625) (2018)
5. A. Totin, E. MacDonald, B. Conner, Additive manufacturing for aerospace maintenance and sustainment. DSIAC J. **6**(2) (2019)
6. ISO/ASTM52900-15, *Standard Terminology for Additive Manufacturing – General Principles – Terminology* (ASTM International, West Conshohocken, 2015). www.astm.org
7. K.M.B. Taminger, J.K. Watson, R.A. Hafley, D.D. Petersen, Solid Freeform Fabrication Apparatus and Method. (Patent # 7,168,935) (2007)
8. K.M.B. Taminger, W. Hofmeister, R.A. Hafley, Use of Beam Deflection to Control Electron Beam Wire Deposition Processes. (Patent # 8,344,281) (2013)
9. K.M.B. Taminger, R A. Hafley, R.E. Martin, W. Hofmeister, W.J. Seufzer, Closed Loop Process Control for Electron Beam Freeform Fabrication and Deposition Processes. (Patent # 8,452,073) (2013)
10. W.A. Seufzer, R.A. Hafley, Height Control and Deposition Measurement for the Electron Beam Free Form Fabrication (EBF[3]) Process. (Patent # 9,764,415) (2017)

11. R.A. Hafley, K.M.B. Taminger, R.K. Bird, Electron Beam Freeform Fabrication in the Space Environment. Proceedings of the 45th AIAA Aerosciences, Reston, VA, (pp. 13879-13887) (2007)
12. R.E. Honig, Vapor pressure data for the solid and liquid elements. RCA Rev. **30**, 285–305 (1969)
13. S.N. Sankaran, R.A. Hafley, C.L. Lach, K.M. Taminger, A Microstructural and Microanalytical Study of the Effect of Processing Parameters on the Aluminum Loss and Deposition Efficiency of EBF³ Ti-6Al-4V Alloys. Presented at the 19th AeroMat Conference and Exposition, Austin, TX (2008)
14. Y.M. Yaman, M.C. Kushan, Hot cracking susceptibilities in the heat-affected zone of electron beam-welded Inconel® 718. J. Mater. Sci. Lett. **17**(14), 1231–1234 (1998). https://doi.org/1 0.1023/A:1006514431915
15. D.L. Ellis, H.R. Gray, M. Nathal, *Aerospace Structural Materials Handbook Supplement GRCop-84* (2001)
16. B.K. Stanford, C.V. Jutte, Comparison of Curvilinear Stiffeners and Tow Steered Composites for Aeroelastic Tailoring of Transports. Proceedings of the 34th AIAA Applied Aerodynamics Conference, Washington, D.C., (paper no. AIAA 2016-3415) (2016)
17. P. Joshi, S.B. Mulani, R.K. Kapania, Experimental validation of the EBF3PanelOpt vibroacoustic analysis of stiffened panels. J. Aircr. **52**, 1481 (2015, January 20). https://doi.org/10.2514/1.C032982
18. S. Lin, E. Hoffman, M. Domack. Distortion and Residual Stress Control in Integrally Stiffened Structure Produced by Direct Metal Deposition. Presented at the 18th AeroMat Conference and Exposition, Baltimore, MD (2007)
19. M.S. Domack, K.M.B. Taminger, M. Begley, Metallurgical mechanisms controlling mechanical properties of aluminum alloy 2219 produced by electron beam freeform fabrication. Mater. Sci. Forum **519–521**, 1291–1296 (2006)
20. C. Brice, R. Shenoy, M. Kral, K. Buchannan, Precipitation behavior of aluminum alloy 2139 fabricated using additive manufacturing. Mater. Sci. Eng. A **648**, 9–14 (2015)

Chapter 9
Advanced Characterization of Multifunctional Nanocomposites

Nellie Pestian and Dhriti Nepal

9.1 Introduction

MNCs combine structural and functional properties, where the structural aspect refers to mechanical performance, including increases in modulus, strength, toughness, and resistance to fatigue. Functional properties include electrical, optical, thermal, and magnetic susceptibility [1].

Two key challenges associated with these nanocomposites include achieving uniform dispersion of nanoparticles in a polymer matrix and controlling interfacial interactions between the nanofiller and the matrix [2–4]. If sufficient dispersion is not attained, nanoparticles may aggregate and create sites of stress concentration. Weak interfacial interactions, resulting from incompatible surface chemistries and few covalent bonds between filler and matrix, induce chain slippage, bond breaking, and void creation, which all contribute to crack initiation and propagation under a mechanical load. Thus, mechanical weak points at the nanoscale contribute to failure at the bulk scale. After decades of research on polymer nanocomposite synthesis, researchers have developed techniques implementing particle surface treatment (using appropriate surfactants, polymers, or coupling agents) and effective mechanical mixing to uniformly disperse nanoparticles in a polymer matrix [4]. Two methods are being used for improvement of the interaction between nanofiller and polymer. The first involves physical wrapping of polymer chains on the nanofiller via secondary forces such as electrostatic, steric, and van der Waals forces; hydrogen bonding; and Lewis acid-base interactions. With this technique, the physically adsorbed polymer or oligomer chains minimize voids and enhance the mechanical performance. The second method encourages covalent bonding between the nanofillers and the host polymer matrix by controlling the polymer grafting density of

N. Pestian · D. Nepal (✉)
Materials and Manufacturing Directorate, Air Force Research Laboratory,
Wright-Patterson Air Force Base, OH, USA
e-mail: pestian.3@wright.edu; dhriti.nepal.1@us.af.mil

© Springer Nature Switzerland AG 2020
M. E. Kinsella (ed.), *Women in Aerospace Materials*, Women in Engineering
and Science, https://doi.org/10.1007/978-3-030-40779-7_9

the corona structures. The resulting increase in covalent bonds ensures strong coupling between the filler and particles.

Difficulties with aggregation and weak interfacial interactions depend on the concentration of surface atoms on filler particles. At nanoscale dimensions, the surface-area-to-volume ratio is drastically increased from that of a micron-sized filler (Fig. 9.1). This geometric property influences the mechanical, chemical, and functional properties of the composite, because surface atoms are more active than atoms central to the nanofillers. An increase in relative concentration of active surface atoms causes an increase in the interaction between nanofillers and the surrounding matrix, and this effectively changes the composite strength and toughness, chemical and heat resistance, optical interactions, and more. Different surface functionality, shape, size, and distribution of nanofillers affect the polymer reaction, packing, and polymerization of the matrix close to a filler particle, causing the formation of an interphase region.

9.2 The Interphase

The interphase is a region between a particle and a surrounding matrix, where the chemical and thermomechanical properties are expected to be different than the bulk (Fig. 9.1a). It is often induced as a result of the difference in polymer packing surrounding a nanoparticle due to the difference in surface chemistry, roughness, and thermal conductivity. The thickness of the interphase depends on many complex factors, including the chemistry of the matrix, surface chemistry of the filler, surface roughness and shape of a filler, and polymer processing conditions. One major question posed to the scientific community pertains to the thickness and the local properties of the interphase region, which is very hard to characterize and

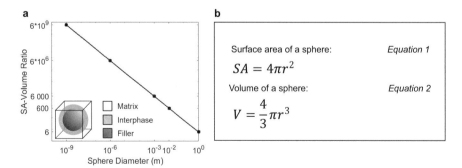

Fig. 9.1 Graphical representations of the exponential increases in surface-area (SA)-to-volume ratio with decreasing particle diameter (assuming a spherical particle) (**a**). A logarithmic scale is used to illustrate exponential nature of the relationship. The mathematical equations used for generation of the graph are given in (**b**). *SA* and *V* abbreviate surface area and volume, respectively. *r* is the radius of the filler particle ($r = d/2$)

quantify. The ability to predict or measure these parameters would greatly assist in the prediction of composite properties. This is because interphase volume fraction of a polymer nanocomposite increases exponentially with an increasing volume fraction of nanoparticles and decreasing size of a nanoparticle.

The mechanical properties of the interphase may be very different from those of the bulk matrix, affecting the ability to transfer mechanical stress from the matrix (higher toughness) to the filler (higher yield strength). Therefore, higher volume percent of interphase increases uncertainty in the overall performance of the composite. Characterization of the chemical bonds and mechanical properties within the interphase could improve understanding of the region; however, characterization at the nanoscale poses a significant challenge. The objective of this chapter is to give perspective on the advanced techniques for nanocomposite characterization and discuss some of the challenges.

9.3   Nanoscale Characterization: Techniques and Challenges

Despite significant progress over the last few decades regarding the synthesis and processing of nanocomposites, the scientific community has achieved only a limited understanding of material properties at the nanoscale. This knowledge gap is caused by a lack of effective characterization tools for nanoscale study of polymer nanocomposites. It is known that surface chemistry of nanoparticles influences the polymer packing and polymer reaction, which can create an interphase zone between a nanoparticle and a matrix. Challenges and limitations that already exist for microscale characterization are compounded when attempting to study composites at the nanoscale. The following sections will briefly introduce three instruments used for micro- and nanoscale characterization, giving detail about the applications and limitations of each instrument in the context of nanocomposite study. Methods of using different instruments to investigate and overcome the challenges of each other are also discussed.

9.4   AFM of Polymer Nanocomposites

One powerful tool for nanoscale characterization is atomic force microscopy (AFM), which is a type of scanning probe microscopy (SPM). AFM can be used to obtain an abundance of information about a nanocomposite sample, including measurements of surface roughness [5], filler dispersion [6], and material stiffness (elastic modulus) [5, 7, 8]. The AFM also generates images of various properties, including physical morphology [5–7, 9]; chemical adhesion [7, 10], surface energy [11, 12], and work function [11, 12]; mechanical modulus [7, 8] and phase [13]; and magnetic [14, 15] and electrical [9, 12] properties. As interest in the nanocomposite interphase region moves to the forefront of composite study, the nanoscale resolution provided by AFM is recognized as an invaluable asset. Groundbreaking achievements have been made in the field of nanocomposite characterization, specifically measurements of nanofiller modulus and thickness [7, 8], studies of variations in properties across the interphase [5, 7, 8], and measurements of interphase thickness [5, 7, 8, 13].

The AFM is comprised of a piezoelectric scanner which moves the sample under the tip of a very sharp cantilever (probe) in the manner shown in Fig. 9.2a. Early methods of nanocomposite characterization used force modulation and phase mapping methods for evaluation of mechanical properties [5, 13]. Such techniques were capable of mapping local variations in stiffness, and elastic modulus data could be approximated with post-processing, but a newer approach called PeakForce Quantitative Nanomechanical Mapping (PF-QNM) adds the capability of mapping quantitative elastic modulus data in real time [16]. PF-QNM works by scanning the AFM cantilever over the sample surface and using the tip to exert a set force on the sample at a certain frequency. Data from these sampling points correspond to pixels in the generated image. A force vs. penetration (force-indentation) curve is generated for every sampling point and gives information about local properties such as the height, deformation, and adhesion of the sample.

An exemplary pair (approach and withdrawal) of force-indentation curves is provided in Fig. 9.2b, and the numbered regions of the curves are explained as follows. The probe approaches the sample without encountering any force during region (1) of the curve. At point (2), the probe "snaps" to the sample surface due to attractive forces between the tip and the sample; this is interpreted as a negative force. The moment after this phenomenon when the force on the probe is zero is labeled as the contact point and is recorded as the height of the sample at that location. The probe tip then applies force to indent the sample until it reaches the peak force setpoint at

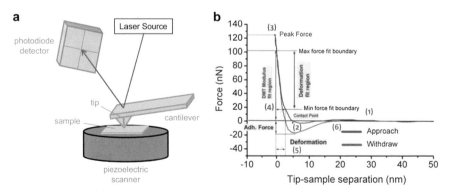

Fig. 9.2 Schematic of AFM function (**a**), including an exemplary force-indentation curve (**b**). Such a force curve is acquired at every pixel of a PF-QNM scan, and mechanical properties such as deformation, DMT modulus, and adhesion are mapped based on the measurements shown. (**b** Adapted with permission from Lorenzoni et al., *Assessing the Local Nanomechanical Properties of Self-Assembled Block Copolymer Thin Films by Peak Force Tapping.* Copyright 2015 American Chemical Society)

(3), after which it begins to withdraw through region (4). This linear region of the withdrawal curve is used, along with knowledge of the cantilever spring constant and tip radius, to calculate the local elastic modulus according to the Derjaguin-Muller-Toporov (DMT) model of surface contact [17]. There is a similar, yet larger, negative dip in the force curve on the withdrawal (compared to the approach), which occurs when the probe tip must overcome adhesive forces as it pulls off of the sample surface (5); the difference between zero and the minimum force in this region is recorded as the adhesive force of the sample. Additionally, the hysteresis area between the approach and withdrawal curves can be used to calculate energy dissipation of one tap. Finally, the curve flattens out once more at zero force (6), signifying that the probe tip is no longer in contact with the sample. The measured and calculated properties (height, modulus, and adhesion) are mapped spatially to create images of the scan area, such as those shown in Fig. 9.3. In determining the appropriate scan size and pixel dimension for these images, it is important to consider the resolution of the AFM.

Resolution, defined in optical microscopy as the smallest distance between two distinctly identifiable objects, is vital to any nanoscale study. In some cases, the resolution limit of an instrument may be too high to conduct a meaningful study of a given length scale, so it is critical to be aware of that lower bound. AFM resolution is limited primarily by the size and geometry of the cantilever tip in comparison with the size and geometry of features to be scanned [18]. For this reason, resolution down to 3–5 nm is theoretically possible; however, a range of 10–20 nm resolution may be more realistic when considering how contact with the sample wears down a very sharp tip over time. The tip diameter is checked periodically using certain calibration steps, and a used tip is replaced once the diameter becomes too large to maintain useful resolution.

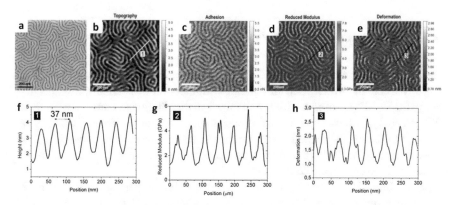

Fig. 9.3 SEM micrograph (**a**) of a block copolymer, compared to AFM images mapping height (**b**), adhesion (**c**), DMT modulus (**d**), and deformation (**e**). AFM images of mechanical properties were obtained using PF-QNM. Cross sections of data from the height, modulus, and deformation images are plotted in (**f**), (**g**), and (**h**), respectively. (Adapted with permission from Lorenzoni et al., *Assessing the Local Nanomechanical Properties of Self-Assembled Block Copolymer Thin Films by Peak Force Tapping*. Copyright 2015 American Chemical Society)

Unfortunately, a trade-off exists between obtaining images with excellent resolution and accurately representing the elastic modulus. The DMT theory is based on contact between a sphere and a plane; therefore, using a sharply pointed AFM tip frequently overestimates the elastic modulus (especially for soft materials such as epoxy) due to the model's neglect of stress concentrations that arise during indentation [19]. With a sharp tip, the approximated value for local modulus does not approach the nominal material value until a relatively deep penetration depth is attained [19]; unfortunately, deeper penetration depths are also more likely to cause plastic deformation in epoxies. In order to find a true elastic modulus, deformation should be fully elastic. For this reason, PF-QNM is preferred over a nanoindenter when studying epoxy nanocomposites, as PF-QNM is capable of operating within the range of smaller indentation forces required by the epoxy [19]. Cantilever tapping parameters such as the amplitude, frequency, and force setpoint should be carefully adjusted during initial calibration to avoid plastic deformation.

Plastic deformation may be observed by abnormal image patterns in successive AFM scans of the same area, but the phenomenon is best visualized with scanning electron microscopy (SEM). Despite artifacts inherent to electron collection, such as inconsistent contrast, SEM eliminates the sort of artifacts that are caused by incompatible tip and sample geometries in AFM to show true morphology of a sample. One example of this is imaging of nanocomposites with 2D nanoplatelets such as graphene as the reinforcing phase. Some platelets may protrude from the sample surface after fracturing or microtoming, and the tapping AFM cantilever can crush these platelets onto the epoxy surface, skewing height and modulus measurements. A non-contact technique like SEM leaves the platelet intact. AFM gives quantitative values of height, while SEM provides qualitative verification for these values.

9.5   SEM of Polymer Nanocomposites

Besides correlation with AFM, SEM has proven its usefulness for nanocomposite study, particularly when used in parallel with electron-dispersive x-ray spectroscopy (EDS). SEM is used to observe the size and distribution of nanofillers [6, 9, 12, 20] and analyze crack initiation and propagation on fracture surfaces [20]. EDS can be used to distinguish nanofillers from the matrix and inspect any chemical functionalization of nanofillers through semiquantitative examination of the elemental makeup of the composite constituents.

SEM, like AFM, offers nanoscale resolution with many parameters to consider for optimal imaging. An SEM operates by rastering an incident electron beam over a region of the sample. A typical schematic of the microscope is shown in Fig. 9.4a. The beam is focused and redirected throughout its raster pattern by a series of electromagnetic coils, or lenses, that are arranged between the electron gun and the sample. Incident electrons from the beam are scattered by the sample, either

Fig. 9.4 Schematic of SEM layout (**a**), including some sample SEM images. (**b**), (**c**), and (**d**) depict fracture surfaces of a neat epoxy (**b**) as well as crack initiation (**c**) and fast-fracture (**d**) sites of an exfoliated clay nanocomposite. (**e**), (**f**), and (**g**) illustrate some common challenges in SEM imaging. (**e**) demonstrates the unrealistic enhancement of contrast on steep edges of topographical features. (**f**) was acquired at low magnification after acquiring (**e**), depicting carbon contamination on the scan area after imaging. (**g**) demonstrates the streaking and glowing of a nonconductive particle under an electron beam. (**b**–**d** Adapted with permission from Wang et al., *Epoxy Nanocomposites with Highly Exfoliated Clay: Mechanical Properties and Fracture Mechanisms.* Copyright 2005 American Chemical Society. Scale bars: 20 um for (**b**, **d**), 5 um for (**c**, **f**, **g**), and 1 um for (**e**))

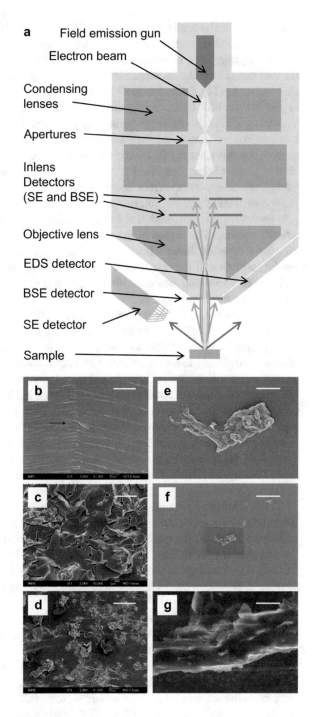

elastically or inelastically, and then collected by a detector. Once incident electrons have been scattered by the sample, they are known as signal electrons.

The electron beam is accelerated by some voltage; this voltage determines how far into the sample the incident electrons penetrate and from what depth they are able to escape [21]. Backscattered electrons (BSEs) are elastically scattered at large angles, so the BSE detector is positioned in line with the incident electron beam. Secondary electrons (SEs) are inelastically scattered, meaning that incident electrons transfer a portion of their energy to allow electrons within the sample to escape. For this reason, SEs have much lower energy than BSEs, and they are only able to escape when generated at the surface of the sample. BSEs, with higher energy, are able to escape from deeper penetration depths; however, this capability is dependent on the atomic number of the sample constituents. Therefore, SE images give topographical information about the sample, while BSE images show compositional contrast. The region of the sample from which electrons of different energies can escape is called the interaction volume, shown in Fig. 9.5e, and it limits the spatial resolution of the micrograph. The interaction volume expands in depth and width with increasing beam acceleration voltage; therefore, better resolution is attained by using lower accelerating voltages.

Perhaps the greatest challenge associated with SEM of polymeric materials is the need for an electrically conducting sample (due to the use of an electron beam), since polymers are generally nonconductive. Charge builds up on the surface of a nonconductive sample and interferes with both the incident and scattered electrons to create distortions and artifacts in the final image such as streaking or glowing [21]. Some examples of resultant images are given in Fig. 9.4e–g. One particularly problematic result from this buildup of charge is the appearance of the sample moving, or drifting, while scanning is taking place. When scanning a very small area (i.e., micron dimensions or less) at high magnifications and considering the slow scan speed required for high-resolution imaging, even a very small drift velocity can cause the area of interest to move out of the scan area within the time required to acquire an image.

The first solution for mitigating charging is the same as that for high-resolution imaging: reduce accelerating voltage. Unfortunately, excessively low acceleration voltage also reduces the amount of signal electrons able to escape for collection by the detector. Resulting images have poor contrast and poor signal-to-noise ratios, and artifacts from sample charging still occur to some degree. A few methods exist to work in parallel with decreasing accelerating voltage to yield high-quality SEM micrographs.

One common and effective solution to reduce sample charging artifacts is to sputter coat the sample with a thin layer of a conducting metal such as gold, palladium, or iridium [21]. This layer is typically no more than 10 nm thick and may need to be thinner if fine morphological detail or EDS elemental analysis is required. Each type of coating has different benefits and drawbacks; however the iridium (Ir) has proven to coat the sample with a more uniform thickness and therefore performs well for very thin layers down to 2 nm. This conductive layer, if reasonably uniform,

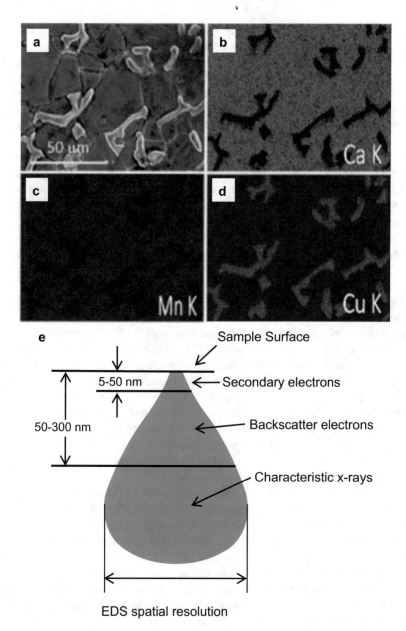

Fig. 9.5 Secondary electron image (**a**) of an oxide ceramic, including EDS map images showing areas of Ca (**b**), Mn (**c**), and Cu (**d**) concentration. EDS mapping has an inherently lower resolution than secondary electron imaging, because the x-rays detected for EDS mapping are generated from the deepest and widest regions of the interaction volume, as shown in (**e**). (**a–d** Adapted with permission from Song et al., *Grain Boundary Phase Segregation for Dramatic Improvement of the Thermoelectric Performance of Oxide Ceramics*. Copyright 2018 American Chemical Society)

largely prevents the buildup of charge at any one location on the sample surface so long as there is a conducting path to ground.

Conductive coating reduces streaking effects in micrographs due to charging, but it is not always perfectly effective for reducing drift. A technique called stage biasing is ideal for this purpose. Stage biasing is often used as a way of improving resolution and contrast in SEM micrographs; however, it also has demonstrated success in reducing drift artifacts for nonconductive samples. The principle behind the method is to decrease the spot size of the beam and therefore improve spatial resolution, by decelerating the beam with both the objective lens and a bias applied to the sample holder [22]. This bias accelerates signal electrons from the sample surface to increase signal at low accelerating voltages and prevent charge buildup, and it induces an electric field which alters the paths of the signal electrons to increase contrast, diminish charging effects, and maintain sample stability.

Hence, high-resolution and high-magnification SEM of a polymer nanocomposite is possible with careful sample preparation and adjustment of scanning parameters. This allows for use of EDS to confirm the presence of filler particles within the matrix for correlation with SPM images. EDS involves focusing the electron beam at a point on the sample and collecting a spectrum for that point. Interpretation of the spectrum gives information about the chemical elements present in the sample. During spectrum collection, incident electrons from the SEM beam displace inner shell electrons in the atoms of the sample [23]. Higher-energy electrons in the atom then relax to this vacancy in the lower energy level, giving off x-rays with specific energy. These x-rays are collected by a separate detector and sorted into a histogram according to their energy; this histogram is presented as a spectrum of counts vs. energy. The energies of x-rays given off by relaxations from each shell of each element are known and can be looked up in tables. Therefore, through analysis of an EDS spectrum, the presence of specific elements can be confirmed. For special cases, care is taken so that the spectra can be analyzed quantitatively to determine chemical composition through ratios of elements.

The SEM can also map the prevalence of elements of interest by acquiring such a spectrum at each pixel of an image. (Some examples of these maps are shown in Fig. 9.5.) The spatial resolution for these images is not quite as good that of SE images, because characteristic x-rays come from deeper and wider regions of the interaction volume as shown in Fig. 9.5e [23]. These map images are useful for comparison with an electron or AFM image showing morphological detail. However, the electron beam often has to scan over a region numerous times in order to collect enough x-ray signal for a clear and decisive map image. This repeated scanning exacerbates the issue of sample destruction in the SEM.

The destructive nature of SEM/EDS also continues to pose a challenge. The electron beam decomposes hydrocarbons, which are prevalent in polymers, and deposits carbon contamination on the surface of the sample [21]. This contamination problem is exacerbated at high magnifications when the beam is rastering over a smaller area. Decreasing the accelerating voltage can decrease the amount of carbon contamination, but it is almost impossible to entirely avoid damage for most polymer samples. A very level sample surface helps, so that beam focusing and the accompanying damage can be carried out on an area far from the region of interest, but it is important that SEM/EDS be the last method of characterization employed on any given sample area due to its destructive nature.

Another powerful electron microscope technique is transmission electron microscopy (TEM). TEM differs from SEM in that the sample for inspection must be very thin (i.e., 10–100 nm thickness) so that electrons will transmit through the sample before being collected by detectors which are on the opposite side of the sample from the electron gun. TEM provides a detailed view of internal microstructure and defects at atomic resolution. With higher resolution than SEM and additional capabilities including electron energy loss spectroscopy (EELS), compositional analysis, and crystallography, TEM is excellent for nanoscale characterization. Unfortunately, in order for the electron transmission to be successful, TEM requires incredibly high acceleration voltages (~200 keV) which exacerbate the existing issues of charging and carbon contamination for polymer samples. Nevertheless, TEM is widely used in polymer nanocomposite characterization to analyze the size, shape, and distribution of nanofillers, measure crystallographic distances, and examine exfoliated layers of 2D fillers (e.g., graphene).

9.6   Atomic Force Microscopy with Infrared Spectroscopy

While EDS is valuable for elemental analysis, it gives no information about the types, prevalence, and distribution of chemical bonds. Bulk characterization techniques such as Fourier transform infrared spectroscopy (FT-IR) are limited by their relatively large spot size and optical diffraction limits. In order to inspect chemistry on a nanometer scale, another type of SPM uses an indirect measurement method combining atomic force microscopy and infrared spectroscopy (AFM-IR) in the manner shown in Fig. 9.6 [24]. Initial development of this technique was based on contact mode AFM, but newer versions include PF-QNM capabilities [25]. Additionally, some variations on the AFM-IR technique include exploration of thermal properties and material stiffness [26].

AFM-IR uses a gold-coated silicon AFM probe to detect rapid thermal expansions that occur when the pulsed IR laser source is tuned to a wavenumber that is absorbed by the sample [24]. As the IR laser is focused on a point and pulsed at a certain wavenumber, thermal expansions occur in time with the pulsing of the laser and excite resonant oscillations in the AFM cantilever tip in contact with the sample surface. The measured amplitudes of the cantilever oscillations relate directly to the IR absorbance for the incident wavenumber. Two methods of data collection arise from this technique: spectrum acquisition at a point and wavenumber mapping over an area. Examples of results from both methods are shown in Fig. 9.6 [25]. The spatial resolution of AFM-IR is slightly larger than that of AFM, because it must account for the width of the thermal expansions. The widths of these thermal expansions are somewhat dependent on both the material and the thickness of the sample, so AFM-IR resolution is difficult to determine exactly.

Thin polymer films with 100–300 nm thickness are ideal for achieving the best spatial resolution and high signal-to-noise ratio (SNR) with AFM-IR. Bulk specimens complicate the mechanism of the thermal expansions, decreasing the SNR of collected spectra. Sections that are too thin struggle to achieve thermal expansions great enough to overcome the SNR, and artifacts might occur due to interactions with the substrate. Since it is not always possible to create such thin films from

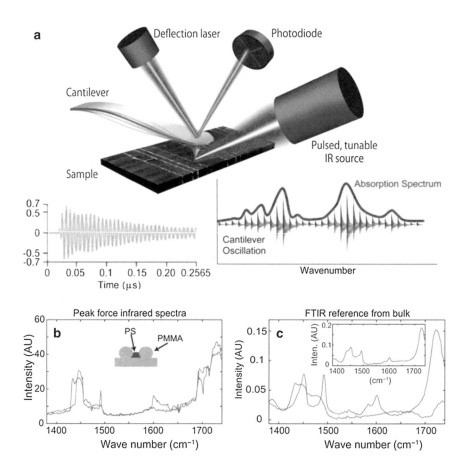

Fig. 9.6 Schematic of AFM-IR function (**a**), with examples of point spectra (**b**) and IR mapping (**e–g**) results from AFM-IR. The blue spectrum in (**b**) corresponds to the polystyrene (PS) domain, indicated by the bottom arrow in the topography image (**d**); the red spectrum is from poly(methyl methacrylate) (PMMA) domain (top arrow). FT-IR spectra of bulk PS (blue), bulk PMMA (red), and the bulk block copolymer (inset) are provided for comparison (**c**). AFM-IR was combined with PF-QNM for this study to also generate topography (**d**), modulus (**h**), and adhesion (**i**) maps of the exact scan area. (**a** Adapted with permission from Dazzi and Prater, *AFM-IR: Technology and Applications in Nanoscale Infrared Spectroscopy and Chemical Imaging.* Copyright 2017 American Chemical Society; **b–i** Adapted with permission from Wang et al., *Nanoscale simultaneous chemical and mechanical imaging* via *peak force infrared microscopy.* Copyright 2017 American Chemical Society)

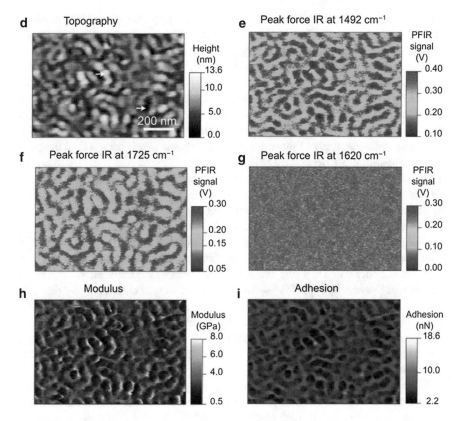

Fig. 9.6 (continued)

typical spin casting methods for thermoset polymers or nanocomposite samples, films are typically cut to the desired thickness using microtomy and floated onto substrate. Best results are obtained when silicon substrates are coated with gold because gold does not show absorbance in the mid-IR region of interest, so there is little risk of substrate interference in the spectra.

To acquire a local chemical spectrum at a point, the cantilever remains positioned over a single sampling location, and the laser is tuned through a range of wavenumbers. For example, 1800–900 cm^{-1} includes the fingerprint region and the peaks at 1500 and 1600 cm^{-1} that are commonly seen in epoxy samples. These local spectra are comparable to data acquired using FT-IR. Parameters set for spectrum acquisition include the starting and ending wavenumber, the spectral resolution, and the co-averages. The spectral resolution can be set in terms of either number of total data points acquired over the defined range of wavenumbers or number of cm^{-1} per point. The co-average value determines the number of measurements averaged to generate a data point for the spectrum; this is typically set to 128 or 256 for best results. There is also an option for enabling multi-region capability, which scans multiple ranges of wavenumbers within the mid-IR spectra using different laser powers for each region. The software built with AFM usually has the capability to acquire multiple spectra automatically and sequentially at different user-defined locations. The locations can be defined randomly in space or arranged in a line or rectangular array.

After spectrum acquisition at multiple locations, a specific wavenumber may be identified that varies spatially. To better understand this spatial variation, an IR map is acquired over the scanning area by tuning the IR laser to the wavenumber of interest and sampling an array of locations that will become pixels for the final image. A few parameters must be optimized for accurate representation with IR mapping. The scanning frequency is defined in Hz, to determine the time for one trace and retrace of the same line of pixels by the cantilever. The X- and Y- resolutions are set to determine the number of pixels generated per line and number of lines in the image, respectively. A separate co-average value is defined for mapping, which is the number of samples averaged to generate a pixel in the IR map. Note that this co-average value is often significantly less than that for spectrum acquisition, typically set at either 8 or 16. There is a necessary compromise between these parameters to avoid oversampling; if the software attempts to read data too many times for one pixel, while the cantilever is scanning at excessively high frequency, the cantilever will move to the location of the next pixel while still reporting data to the previous pixel. The resulting image fails to show correct data for each pixel. Thus, the scanning frequency, X-resolution, and co-average values must be optimized to get the best and most accurate image.

When using technique that relies on measuring height differences from thermal expansions, there is an obvious challenge associated with mapping IR absorbance over an area with raised surface features. When the cantilever experiences an increase in height, it records an increase in cantilever deflection which is interpreted as greater IR absorbance in the IR map. Therefore, the tallest physical point on a sample is often measured as the location of greatest IR absorbance for the region, regardless of the local chemistry. Sample height is recorded in nm, while IR absorbance is recorded as deflection volts, and there is no direct correlation to simply "subtract" the height artifacts. A useful method for interpreting data free of height artifacts involves calculating the pixel-by-pixel ratio between an image acquired at a wavenumber of interest and another image acquired for a reference wavenumber. This reference wavenumber should be one that demonstrates baseline IR absorbance in a spectrum. Thus, the relative absorbance of multiple wavenumbers can be compared free of height artifacts, when referenced to the absorbance of a common wavenumber.

Another notable challenge in data acquisition and interpretation using AFM-IR arises from thermal drift, a phenomenon common among SPM instruments. Thermal drift is caused by thermal expansion of the piezoelectric material that controls the scanning of the AFM cantilever [27]. The piezoelectric controls AFM scanning with mechanical expansions and contractions due to applied voltage differences, so the addition of thermal expansions introduces a spatial uncertainty to an AFM image [27]. Assuming approximately linear thermal drift, there is a constant drift velocity which affects image accuracy to an increasing degree with decreasing scan area and scan rate. SEM of a sample area after AFM scanning reveals the significance of drift. The drift phenomenon further complicates the process of AFM-IR scanning in a few ways. When collecting data for a point spectrum, the cantilever gradually drifts away from the point of interest with long acquisition times, which introduces uncertainty. Furthermore, drifting complicates the process of correlating successive images for the removal of height artifacts because there is no guarantee that the pixels of each scan perfectly overlap and correspond to the same physical location.

Correction of this drift, both post-processing and during scanning, is of great interest as length scales continue to decrease. Faster acquisition times are key for mitigating drift, and recently success has been demonstrated with a new instrument that combines PF-QNM and AFM-IR for improved speed and spatial resolution [25]. For current AFM-IR systems, temperature control is the simplest method of counteracting drift during scanning. Lower temperatures cause less thermal expansion and therefore less drift. There are also relatively straightforward post-processing calculations that can correct drift within an image for more accurate representation and can correct drift between subsequent images to allow for image ratio calculation. The methods involve one-dimensional normalized cross correlation of each line of two successive AFM scans of the "same" area. Cross correlation determines the offset of a feature between two scans, and the relative locations can be averaged to center the feature in its true location. When creating one cohesive height image, any overlapping pixel values are averaged. If the two images are IR absorbance maps intended for calculating a ratio image, the offset is determined using the corresponding height images, and a ratio is found between pixels of the IR maps.

9.7 Summary: The Power of Multiple Techniques

As interest in MNCs continues to grow and new developments are made for aerospace applications, it becomes increasingly important to ensure a complete understanding of their properties in which the damage and failure mechanisms of materials are of significant concern. In particular, there is a strong interest in developing high-fidelity computational predictive tools for performance prediction; it is critical to parameterize computational models with experimentally identified parameters. In order to understand the damage and failure mechanism of nanocomposites, characterization must take place on multiple length scales. Multiscale characterization demands a variety of techniques, including optical and electron microscopy; AFM and AFM-IR; x-ray computed tomography; x-ray diffraction; FT-IR, Raman, and UV-Vis spectroscopy; differential scanning calorimetry; and dynamic mechanical analysis. Significant advances have been made in the last decade, particularly with the development of new forms of SPM (e.g., PF-QNM and AFM-IR), which probe elastic modulus and chemical bond type at nanoscale resolutions. Many challenges are left to overcome in the field of nanoscale characterization, but the issues may be studied, improved upon, and even overcome through the combination of multiple techniques. The lack of mechanical contact while imaging in an SEM makes it a perfect tool for studying scanning artifacts in AFM images, while PF-QNM and AFM-IR give chemical and mechanical information about a sample that cannot be obtained with an SEM. By developing innovative methods of corroboration between various characterization tools with nanoscale resolution, the properties of MNCs can be studied and understood for new and exciting applications.

References

1. J. Baur, E. Silverman, Challenges and opportunities in multifunctional nanocomposite structures for aerospace applications. MRS Bull. **32**(4), 328–334 (2011)
2. K.I. Winey, R.A. Vaia, Polymer nanocomposites. MRS Bull. **32**(4), 314–319 (2007)
3. S.K. Kumar et al., 50th anniversary perspective: Are polymer nanocomposites practical for applications? Macromolecules **50**(3), 714–731 (2017)
4. H.B. Gu et al., An overview of multifunctional epoxy nanocomposites. J. Mater. Chem. C **4**(25), 5890–5906 (2016)
5. T.D. Downing et al., Determining the interphase thickness and properties in polymer matrix composites using phase imaging atomic force microscopy and nanoindentation. J. Adhes. Sci. Technol. **14**(14), 1801–1812 (2000)
6. T. McNally et al., Polyethylene multiwalled carbon nanotube composites. Polymer **46**(19), 8222–8232 (2005)
7. A. Pakzad, J. Simonsen, R.S. Yassar, Gradient of nanomechanical properties in the interphase of cellulose nanocrystal composites. Compos. Sci. Technol. **72**(2), 314–319 (2012)
8. Y.F. Niu et al., Mechanical mapping of the interphase in carbon fiber reinforced poly(ether-ether-ketone) composites using peak force atomic force microscopy: Interphase shrinkage under coupled ultraviolet and hydro-thermal exposure. Polym. Test. **55**, 257–260 (2016)
9. A. Trionfi et al., Direct imaging of current paths in multiwalled carbon nanofiber polymer nanocomposites using conducting-tip atomic force microscopy. J. Appl. Phys. **104**(8), 6 (2008)
10. X.Y. Li et al., Directly and quantitatively studying the interfacial interaction between SiO2 and elastomer by using peak force AFM. Composites Commun. **7**, 36–41 (2018)
11. L.Q. Guo et al., Electron work functions of ferrite and austenite phases in a duplex stainless steel and their adhesive forces with AFM silicon probe. Sci. Rep. **6**, 7 (2016)
12. C.C. Tseng et al., Surface and mechanical characterization of TaN-Ag nanocomposite thin films. Thin Solid Films **516**(16), 5424–5429 (2008)
13. S.L. Gao, E. Mader, Characterisation of interphase nanoscale property variations in glass fibre reinforced polypropylene and epoxy resin composites. Composites Part a-Appl. Sci. Manufact. **33**(4), 559–576 (2002)
14. A. McDannald et al., Switchable 3-0 magnetoelectric nanocomposite thin film with high coupling. Nanoscale **9**(9), 3246–3251 (2017)
15. H. Wang et al., Microstructure evolution, magnetic domain structures, and magnetic properties of Co-C nanocomposite films prepared by pulsed-filtered vacuum arc deposition. J. Appl. Phys. **88**(4), 2063–2067 (2000)
16. B.N.S. Division, *Quantitative Mechanical Property Mapping at the Nanoscale with PeakForce QNM*. Bruker Corporation, 2012 (Application Note #128)
17. B.V. Derjaguin, V.M. Muller, Y.P. Toporov, Effect of contact deformations on the adhesion of particles. Prog. Surf. Sci. **45**(1–4), 131–143 (1994)
18. V.V. Tsukruk, S. Singamaneni, Basics of atomic force microscopy studies of soft matter, in *Scanning Probe Microscopy of Soft Matter*, Wiley-VCH Verlag GmbH & Co. KGaA. https://onlinelibrary.wiley.com/doi/abs/10.1002/9783527639953.fmatter (2012)

19. M.E. Dokukin, I. Sokolov, Quantitative mapping of the elastic modulus of soft materials with HarmoniX and peak force QNM AFM modes. Langmuir **28**(46), 16060–16071 (2012)
20. K. Wang et al., Epoxy nanocomposites with highly exfoliated clay: Mechanical properties and fracture mechanisms. Macromolecules **38**(3), 788–800 (2005)
21. Y. Leng, Scanning electron microscopy. Mater. Charact. (2013)
22. F. Zhou, L. Han, *Surface sensitive imaging with a bias on the sample* (Carl Ziess Microscopy GmbH, Germany, 2016)
23. Y. Leng, X-ray spectroscopy for elemental analysis. Mater. Charact. (2013)
24. A. Dazzi, C.B. Prater, AFM-IR: technology and applications in nanoscale infrared spectroscopy and chemical imaging. Chem. Rev. (2016)
25. L. Wang et al., Nanoscale simultaneous chemical and mechanical imaging via peak force infrared microscopy. Sci. Adv. **3**(6), 11 (2017)
26. N. Li, L.S. Taylor, Nanoscale infrared, thermal, and mechanical characterization of telaprevir-polymer miscibility in amorphous solid dispersions prepared by solvent evaporation. Mol. Pharm. **13**(3), 1123–1136 (2016)
27. M.S. Rana, H.R. Pota, I.R. Petersen, A survey of methods used to control piezoelectric tube scanners in high-speed AFM imaging. Asian J. Control **20**(4), 1379–1399 (2018)

Chapter 10
Materials and Process Development of Aerospace Polymer Matrix Composites

Sandi G. Miller

10.1 Introduction

This chapter will detail studies in both materials development and process development for polymer matrix composites (PMCs). The two topics differ in that materials development aims to create novel materials for infusion into a given application based on performance requirements. Process development aims to optimize part fabrication for a given platform. PMCs are composed of a reinforcement phase held together by a polymeric matrix. Several material options are available for the reinforcement phase; however, the work described herein will focus on continuous carbon fiber due to its high strength and stiffness. The high strength to weight ratio is attractive in that it enables the manufacture of lightweight load-bearing structures. A thermosetting epoxy resin matrix is often selected for aerospace composite structures due to a broad range of chemistries and processing options available, leading to a range of attainable mechanical and thermal performance. Properties of the matrix material dominate composite mechanical performance in compression and shear, and the temperature capability of the matrix drives the thermal limitations of the composite. Even within the specific material classes described (carbon fiber and epoxy), countless variations are available in terms of constituent specifications and composite performance.

Processes for PMC fabrication are guided by the structural requirements, overall size, and dimensional complexity of the part. Often, sheets of carbon fiber embedded with uncured epoxy (prepreg) are layered to meet a required thickness and then consolidated and cured in an autoclave. As demand for large composite structures has increased, the need to cure outside of an autoclave has grown, and, correspond-

S. G. Miller (✉)
Materials and Structures Division, Ceramic and Polymer Matrix Composites Branch, NASA
Glenn Research Center, Cleveland, OH, USA
e-mail: Sandi.G.Miller@nasa.gov

© Springer Nature Switzerland AG 2020
M. E. Kinsella (ed.), *Women in Aerospace Materials*, Women in Engineering and Science, https://doi.org/10.1007/978-3-030-40779-7_10

ingly, an increasing selection of composite systems that can be consolidated without autoclave pressure has been developed. Alternative to prepreg systems, resin transfer molding (RTM) describes a process where dry carbon fiber preforms are stacked in a closed mold and low viscosity resin is infiltrated through the material.

The following sections detail three specific contributions to NASA composite programs. They include the evaluation of process-related chemistry including cure kinetics, rheology, residual stress development, and heat transfer as relevant to out-of-autoclave processing and the fabrication of thick composite parts. The final section of this chapter summarizes work in a materials development program aimed to improve the aerodynamic efficiency of composite fan blades while maintaining damage tolerance requirements.

10.2   Materials Process Development

10.2.1   Influence of Out-Time on the Out-of-Autoclave Processability of Large Composite Structures [1, 2]

In 2008, the NASA Exploration Systems Mission Directorate initiated an Advanced Composite Technology Project through the Exploration Technology Development Program to support the polymer composite requirements for future heavy-lift launch architectures, specifically the Ares V heavy-lift launch vehicle (Fig. 10.1). Composite structures for heavy-lift launch vehicles were projected to be the largest composite structures ever fabricated for an aerospace application. Some of these large composite shelled structures were designed to exceed 9 m in diameter and 10 m in length.

Fig. 10.1 Artist rendering of the Ares V Rocket which was planned during the NASA Constellation program

At the time, autoclaves large enough to cure the 9–10 m composite shells were not widely available [1]. The large composite structural applications on Ares V inspired the evaluation of autoclave and out-of-autoclave (OOA) composite materials [2].

Manufacture of increasingly large composite components can challenge the room temperature out-life limits of carbon fiber/epoxy prepreg. Prior to cure, most epoxy matrices have a specified out-life, or room temperature shelf life, ranging from 20 to 30 days. During this time, the material slowly cross-links, impacting the processability and ultimate performance of the laminate. The large Ares V composite part specified for out-of-autoclave processing would require approximately 30 days to manufacture. Therefore, it was essential to compare prepreg specified for autoclave and OOA manufacturing, in terms of processing and property retention following an extended out-time. My role specifically included chemical characterization of the resin as it aged at room temperature. The neat resin was characterized by differential scanning calorimetry (DSC) to evaluate cure behavior as a function of out-time. A change in modulus of the uncured prepreg was monitored by dynamic mechanical analysis (DMA) over a 60-day period at room temperature. Composite panels made of the baseline and aged prepreg were also characterized by DMA.

Heat is generated as epoxy cures; therefore DSC provides data in the form of an exothermic peak which is related to thermal energy generated during cure. The data also provide the cure temperature. Variation in the area under the temperature-vs-heat flow curve, from that of a baseline measurement, is indicative of a change in the cure state of the epoxy resin. The area under the cure peak of an epoxy recommended for autoclave processing was measured and is plotted in Figs. 10.2 and 10.3. The data show an initial decrease in the area under the curve after 10 days at

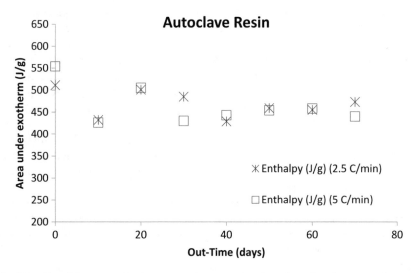

Fig. 10.2 Plot of the area under the cure exotherm as a function of out-time for an autoclave cure resin

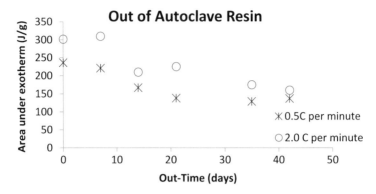

Fig. 10.3 Plot of the area under the cure exotherm as a function of out-time for an OOA cure resin

room temperature. Through the remaining out-time, however, the peak area held steady, up to 70 days at room temperature. This length of time more than doubles the recommended out-time of the resin, and it appears to be stable with respect to cure state, although this is not the only consideration for processability.

Conversely, DSC characterization of the material designed for OOA cure resulted in a steady decrease of the measured area under the exotherm as a function of out-time. The DSC data illustrate a greater susceptibility of the OOA resin to advance at room temperature and dictated a criticality in the manufacturing timeline for composite parts that would be manufactured by the OOA process.

The extended room temperature shelf life of the autoclave-processed materials translated into superior retention of mechanical properties of the cured part. It should be emphasized that in the data below, material was aged 50 percent longer than the manufacturer recommended out-time. The intent was not to choose a material but to bound the usable limits of commonly employed materials within both processing regimes, autoclave and OOA. Figure 10.4 plots the interlaminar shear strength measured from composites manufactured with baseline and aged autoclave prepreg (977-3, 8552-1, 8552-1 FP), as well as baseline and aged OOA prepreg (MTM45-1). Data for aged T40-800/5320 prepreg was not available due to difficulty with panel fabrication.

The chemistry of the resin formulated for autoclave consolidation allows for a longer room temperature workable lifetime relative to the OOA cure material. OOA material that has outstanding performance when used within its shelf life, however, can be unforgiving with respect to out-time.

In summary, resins engineered for autoclave cure exhibited outstanding resilience to room temperature advancement of cure. As a result, composites manufactured using autoclave-recommended prepreg retained mechanical property performance after extended out-times. This was not the case for prepreg designed for OOA processing. However, when used within the vendor-recommended out-time, the composites performed comparably to those cured in the autoclave. For

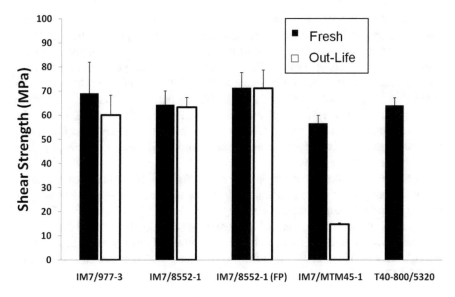

Fig. 10.4 Interlaminar shear strength of composites manufactured from baseline and aged autoclave and OOA prepreg

applications requiring long out-times, improved OOA prepreg processability will be necessary in order to produce high-quality composites with mechanical properties equivalent to or greater than autoclave composites.

10.2.2   Evaluation of Exotherm Development in Processing Thick Composite Parts

NASA has made significant contributions to the design and test of a composite-metallic hybrid gear for rotorcraft gear shafts [3]. In service, such a part would experience high torsional loads. As stated, the structural loads for a given application drive the part design, including thickness, which presents processing challenges with respect to temperature and cure variability in epoxy-based composites. The cure history of an epoxy matrix composite directly influences both thermal and mechanical properties. Therefore, cure conditions are established to provide a uniform degree of cross-linking, void-free consolidation, and minimized residual stress. As panel thickness increases, processing controls over the thermal history become challenging, as the thicker part may experience issues such as thermal spiking and through-thickness temperature gradients, both of which would lead to nonuniform cure and residual stress development [4–6].

Processing concerns are associated with thermosetting composites of appreciable thickness, greater than 2.54 cm, and result from poor ply-to-ply transfer of thermal energy. The predominant sources of heat throughout a cure cycle include oven air convection, heat transfer through the tool, and internal heat generated by the exothermic cure reaction of the resin. Heat generated internal to the part is a particular concern as it is not easily dissipated through the laminate, particularly in the z-direction. In severe instances, this may result in degradation of the material, among the previously mentioned concerns. Therefore, a standard cure cycle may require modification to accommodate the time and temperature requirements of a thick part to properly complete the cure reaction [4].

The thermal component of a typical cure cycle for a thermosetting epoxy includes a thermal ramp from room temperature to a low temperature hold, followed by a second ramp to the cure temperature.

Throughout the initial ramp and low temperature hold, the thermal distribution of a part is limited by the diffusion of heat from surrounding oven air or heat transfer through the tool. Heat transfer as a function of panel thickness can be described by Eq. 10.1 [7]:

$$\rho c \frac{\partial T}{\partial t} = \frac{\partial}{\partial xi}\left(\kappa(T,\phi)\frac{\partial T}{\partial xi}\right)$$

(10.1)

where ρ is the density, c is the specific heat, and κ is the thermal conductivity of the resin.

The temperature increase following the onset of cure is expressed through the following set of equations.

The degree cure as a function of time is expressed in Eq. 10.2:

$$\phi = \frac{H(t)}{H_r}$$

(10.2)

where $H(t)$ is heat generated to time t, H_r is the total heat of reaction, and ϕ is degree of cure.

The rate of change in the degree of cure is expressed through a cure kinetics model, Eq. 10.3:

$$\frac{d\phi}{dt} = \left[A_1 \exp\left(\frac{\Delta E_1}{TR}\right) + A_2 \exp\left(\frac{\Delta E_2}{TR}\right)\phi^m\right](1-\phi)^n$$

(10.3)

where $\frac{d\phi}{dt}$ is the rate of cure, T is temperature, R is the gas constant, ΔE_1 and ΔE_2 are activation energies, and A_1, A_2, m, and n are empirically defined constants from fitting Eq. 10.3 to experimental data.

Complex processing issues arise during the second ramp to cure temperature. At this stage, the rate of resin cure accelerates, and thermal energy due to the exothermic chemical reaction is quickly generated. As a result, the temperature of the matrix is influenced by both heat generated from exothermic chemical reactions and heat conduction from the applied cure cycle. The change in matrix temperature with respect to time is expressed in Eq. 10.4 [7]:

$$\rho c \frac{\partial T}{\partial t} = \frac{\partial}{\partial x_i}\left(\kappa\left(T,\phi\right)\frac{\partial T}{\partial x_i}\right) + \rho H_r \frac{d\phi}{dt} \tag{10.4}$$

The rate of heat transfer through the panel scales with thickness; therefore, slow diffusion of heat generated during the cure of thick panels has the potential to raise internal temperatures to levels significantly greater than the intended cure temperature or the temperature experienced by the surface plies.

In this work, the development of temperature gradients within an increasingly thick part was measured. The influence of a non-uniform temperature distribution throughout the part thickness was evaluated, and ultimately the glass transition temperature (Tg) of coupons from various thickness positions was measured.

Three laminate thicknesses were evaluated in this study and ranged from 0.64 cm (0.25″) to 3.81 mm (1.5″) in thickness. All panels measured 20.3 cm² (8″ × 8″) in the lateral dimensions. Panels were fabricated with axial fibers oriented along the 0° axis (Fig. 10.5). The bagging sequence followed that recommended per the Tencate datasheet [8].

A vacuum-bag-only cure cycle was followed using a programmable benchtop vacuum oven. The oven chamber dimensions were 46 cm × 46 cm × 51 cm (l × w × h) with a convection fan located at the base of the chamber, therefore heating from beneath the panel. The laminate was positioned at the center of the chamber height.

The internal laminate temperature was monitored throughout cure via thermocouples embedded within the laminate thickness. Three thermocouples were inserted during laminate lay-up, each protected by a small non-porous Teflon patch, to allow retrieval and reuse. The thermocouple tip was placed in the direct center of

Fig. 10.5 Sheet of 700S/
TC380 braided carbon
fiber/epoxy prepreg with
predominately ±60° bias
fibers visible

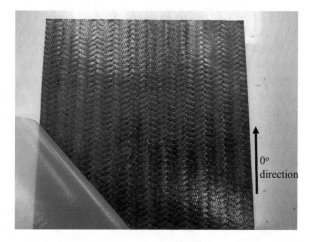

0°
direction

Fig. 10.6 Thermocouples embedded through the part thickness to monitor internal temperature during cure

the specified plies, between the second and third tool-side plies, second and third bag-side plies, and the two central plies. A second panel was fabricated at each thickness with thermocouples embedded between central plies and 25% of the part thickness from the center as shown in Fig. 10.6. Oven air temperature was measured with a thermocouple placed directly above the part. Ultimately, the greatest variation in part temperature was measured between the central and surface plies.

The recommended vacuum-bag-only cure cycle proceeded as follows:

- Ramp from room temperature to 107 °C (225 °F) at a rate of 0.6 °C/min (1 °F/min).
- Hold at 107 °C for 1 hour.
- Ramp to 180 °C (356 °F) at a rate of 0.6 °C/min (1 °F/min).
- Hold for 2 hours.
- Cool at a rate <3 °C/min (5 °F/min) to below 49 °C (120 °F).
- Post-cure at 1 hour at 180 °C. (Half of the coupons produced were analyzed prior to post-cure.)

On processing panels in this study, the ramp rates were set to 1.1 °C/min (2 °F/min). This modification was approved by the manufacturer.

Panels were fabricated to generate coupons for Tg measurement. The cure cycle for these panels followed that of the previous section; however internal part temperature was not monitored. Rather, sheets of non-porous Teflon were inserted to separate the four outermost tool-side and bag-side plies to enable isolation and machining of these specific sections. Similarly, non-porous Teflon was inserted around the four central plies. These sections were de-bulked, as was the entire panel, and following cure, the individual 4-ply thick panel sections were removed and machined for DMA tests. Half of the coupons were post-cured prior to test, and half were tested without post-cure to elucidate variations in thermal characteristics prior to complete resin cure.

Through-thickness thermocouple data were collected for panels of increasing thickness cured on aluminum tooling. Data presented in Fig. 10.7 plot temperature

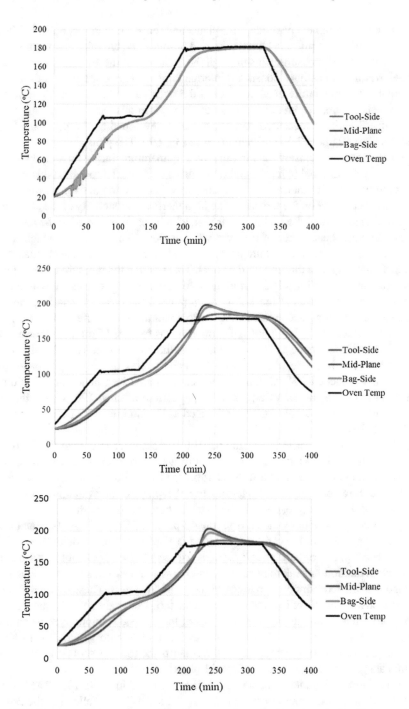

Fig. 10.7 Temperature data generated by thermocouples embedded through the thickness of a 10-ply, 40-ply, and 56-ply panel, respectively

measurements from the bag-side, mid-plane, and tool-side thermocouples in 10-ply (0.64 cm thick), 40-ply (2.54 cm thick), and 56-ply (3.80 cm thick) laminates.

Thermocouple measurements through the thickness of the 10-ply, 0.64-cm-thick panel overlap, indicating a consistent internal laminate temperature. In addition, the internal part temperature does not exceed the oven temperature as a result of heat generated on cure, i.e., an exotherm is not observed in this part.

A distribution in measured part temperature developed as the panel thickness increased to 40 plies (2.54 cm) and 56 plies (3.8 cm). The plots show the tool-side thermocouple most closely tracking the oven temperature during the initial stage of the cure cycle. Internal part temperature is diffusion limited throughout this stage of the cure cycle, with heat transferred to the part via convection of oven air on the bag side or heat transfer on the tool side. The high thermal conductivity of the aluminum tool provided a stable thermal mass through which heat was passed to the laminate. In addition, the heat source of the oven was positioned under the tool which would have accelerated the temperature gains of the aluminum plate and accentuated the temperature disparity throughout the panel. This trend in the measured thermal profile held until approximately 225 minutes into the cure cycle, consistent with the gel time observed by DMA at the 1.1 °C ramp rate. At the onset of gelation, thermocouple data show an accelerated ramp in temperature within the mid-section and bag-side section of the laminate. An exotherm was recorded from both thermocouples for the thicker panels. The temperature at the mid-section exceeded that measured between bag-side plies. Within the center plies of the 56-ply panel, the temperature reached 204 °C. On the bag side, internal heat generation was dissipated through the surface plies, and the temperature reached 190 °C. On the tool side, excess heat accumulation was mitigated as the aluminum tool effectively acts as a heat sink.

Coupons were extracted from the panel thickness to characterize the Tg of the cured laminate. The vendor's cure procedure was used to manufacture these coupons with a final 2-hour hold at 178 °C, followed by a 1-hour post-cure at 178 °C to drive the cure reaction to completion. For the purpose of evaluating cure state throughout the cure cycle, coupons were tested before and after the post-cure.

DMA coupons were extracted from the 40-ply and 56-ply panels. Test data from the 56-ply panel prior to post-cure is plotted in Fig. 10.8. The tan delta curve of the tool-side coupon is of particular interest as we see a reduced peak intensity, a comparative reduction in peak temperature, and a low-temperature shoulder, relative to coupons extracted from other regions of the panel thickness. These thermal characteristics are indicative of incomplete cure at the tool-side region of the panel prior to post-cure. In addition, the storage modulus of the tool-side coupon is reduced relative to that of the bag side or mid-plane as incomplete cure would lead to lower stiffness. Post-cure of the coupons however shifts the tool-side coupon data in line with data generated from bag side or mid-plane.

Data from each panel and coupon location are listed in Table 10.1. The Tg was identified as the maximum of the tan delta peak and is listed following the 2-hour hold at 177 °C and, separately, following an additional 1-hour, free-standing post-cure. The Tg measured for coupons taken from the 40-ply part was consistent

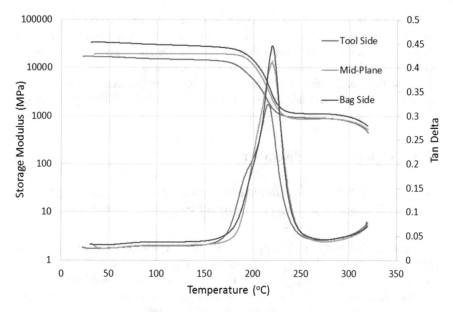

Fig. 10.8 DMA data from the 56-ply panel

throughout the panel thickness and uniformly increased following post-cure. However, a variation in Tg was evident in the 52-ply, 1.5″ part. Prior to post-cure, the tool-side Tg lagged that of the mid-plane by 12 degrees and the bag-side Tg by 9 degrees. The variation in Tg implies a stiffness mismatch would be present within the part. The free-standing post-cure drove Tg of the tool-side material up to match that of the remainder of the parts. Therefore, following the recommended cure profile led to uniform Tg in the material. However, the variation in cure state of the tool-side plies during the cure process is significant as it may lead to residual stress development in curved or complex parts.

In this program, the internal temperature of a series of carbon fiber/epoxy composites was measured throughout the cure profile. As panel thickness increased, internal temperature gradients were measured between the tool-side plies and the remaining bulk of the composite. Effective ramp rates calculated for each section of the part showed an increased ramp rate following gelation for the center portion of the panel. This was attributed to heat generated by the exothermic cure reaction.

Table 10.1 Glass transition temperature data as a function of panel thickness, position within panel, and time within cure cycle

| Location within part | 40-ply | | 52-ply | |
	Tg (°C) 2-h cure on tool	Post-cured (°C)	Tg (°C) 2-h cure on tool	Post-cured (°C)
Tool side	211	216	211	223
Mid-plane	213	217	223	225
Bag plane	213	218	220	224

This heat was dissipated at the tool side due to the high thermal conductivity of the aluminum tool, resulting in the tool-side plies receiving a lower degree of cure relative to the rest of the panel as was evident by Tg measurement. Post-curing the panel equalized the thermal performance throughout the panel thickness. All measured variations in through-thickness thermal and mechanical properties were relatively small; however, they may influence part dimensional stability for complex shapes.

10.3 Materials Development

10.3.1 Thermoplastic Interleave Approach to Meet Damage Tolerance Requirement of a Composite Fan Blade [9]

The primary drivers for technology development within the commercial aircraft community include cost savings and efficiency gains. Common approaches to meet these challenges include weight reduction and increased aerodynamic efficiency of aircraft components. Therefore, polymer matrix composite materials for jet engine fan blades are becoming an attractive alternative to metals, particularly for large engines where significant weight savings can be realized. However, the weight benefit of the composite is offset by a reduction of aerodynamic efficiency, resulting from a necessary increase in blade thickness relative to titanium. The increased thickness is necessary to meet bird strike requirements with the state-of-the-art composite materials. Reduced thickness of state-of-the-art composite blades would translate to structural weight reduction, improved aerodynamic efficiency, and therefore reduced fuel consumption. A materials development effort at NASA Glenn yielded a promising approach to improve composite damage tolerance. This section summarizes the composite materials development, test article design, subcomponent blade leading edge fabrication, test method development, and initial test results from ballistic impact of a gelatin projectile on the leading edge of composite fan blades with improved damage tolerance.

IM7/8551-7 unidirectional prepreg from Hexcel was used for leading edge fabrication. A thermoplastic polyurethane (TPU) veil was incorporated into the leading edge structure to evaluate its influence on composite damage tolerance. The average areal weight of the TPU veil was 15 gsm, and the average fiber diameter was on the nanometer scale. Interlayer veils were chosen for application as they have been shown to improve composite fracture toughness [10].

Fig. 10.9 Leading edge blade cross section with dimensions

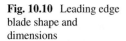

Fig. 10.10 Leading edge
blade shape and
dimensions

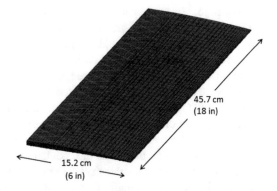

45.7 cm
(18 in)

15.2 cm
(6 in)

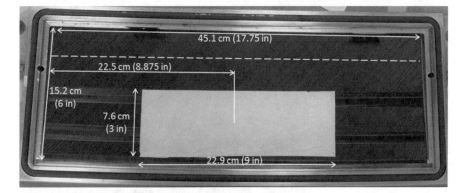

Fig. 10.11 Location of 7.6 cm (3 in.) polyurethane veil layer

The profile of the test blade is shown in Fig. 10.9. It should be noted that the last 3.8 cm (1.5 in) at the mid-blade location was adjusted to be flat and parallel in order to clamp into the test fixture. The profile was then extended in the blade's radial direction (referred to later as the axial direction) to 45.7 cm (18 in) in length, as shown in Fig. 10.10, to create the leading edge blade test article. The profile was designed to compare various material modifications, not to demonstrate an optimum design in either geometry or lay-up configuration. The simplified geometry decreases manufacturing complexity while maintaining a leading edge profile. The changing thickness of the cross section requires ply drops across the blade from 50 plies thick at the base to 2 plies at the leading edge.

Test articles containing three layers of a TPU veil interleave were prepared to evaluate impact resistance imparted by increased interlaminar strain capability of the structure. As seen in Fig. 10.11, the toughening veil was 22.9 cm (9 in) long and placed in the midsection of the length. Two of the layers were 7.6 cm (3 in) wide and the other was 5.1 cm (2 in) wide. The veil placement was chosen to cover a significant portion of ply terminations, thus reducing free edge stresses.

Impact testing was performed in the Ballistics Impact Laboratory at NASA Glenn Research Center to simulate the material response from bird strike at the leading edge. For the impact testing, a single-stage gas gun was used, consisting of a 7 m barrel and a 0.35 m² pressure vessel. The pressure vessel was loaded to a pressure of 1.2 MPa (175 psi), and the pressure was released using a burst disk. The projectile was housed in a cylindrical sabot for protection at the initial pressure release. Both were accelerated through the barrel by the release of pressure. The sabot was halted at the end of the barrel by a sabot arrestor and the projectile continued into the test specimen. High-speed cameras were used to capture the impact of the projectile on the test article and determine the estimated speed. These tests were performed at speeds between 305 and 350 m/s (1000–1150 ft/s). A gelatin bird simulant was used as the projectile, containing ballistic-grade gelatin and microspheres to approximate the density of a bird (~0.9484 g/cm³). The molded projectile was cylindrical, with a length of 7.6 cm (3 in) and diameter of 3.2 cm (1.25 in.).

The projectile velocity and orientation of the leading edge test article were derived from computer simulation of the relative motion of a fan blade and bird during a bird strike event. The blade was clamped along the axial direction at the flat section for cantilevered impact testing. The leading edge specimen and fixture setup for impact testing is shown in Fig. 10.12. The impact location was at the leading edge of the blade where the angle of impact was measured as 66° from the perpendicular to the projectile path at the thickest part of the blade. The blade tip was

Fig. 10.12 Impact testing fixture secured in impact chamber

Fig. 10.13 C006, C007, and C008 after impact

oriented approximately 24° off of the projectile's path. Between the blade and the fixture, there is a 15.2 cm (6 in.) section of clay added to reduce additional pressure loading on the fixed portion of the blade.

Test article C006 and C007 were baseline carbon fiber/epoxy, and the C008 test article included the TPU veil interleave for damage tolerance. Three sheets of the interleave material were placed across the major ply drops to improve interlaminar fracture toughness at these locations. As seen in Fig. 10.13, there was very little visual damage to the C008 test article. No delamination at the leading edge was noted in the interleave region. As with many of the test articles, a small delamination was observed near one tip of the leading edge in an unmodified region due to flexural wave reflection. This failure mode was observed in many of the leading edge tests and is considered an artifact of boundary conditions, occurring after the initial impact.

In this program, the influence of ply configuration and test setup were evaluated such that impact data would reflect modifications to the test article ply structure and materials modifications. An interleave toughening approach was taken in an effort to reduce damage on impact and enable a reduction of composite blade thickness. It was found that incorporation of the lightweight TPU veil provided considerable benefit to the damage tolerance of the composite parts tested in this program.

10.4   Summary

Polymer matrix composites are a class of materials used for primary and secondary structures in both aircraft and spacecraft. The high strength to weight ratio makes them attractive for applications where weight reduction offers a significant benefit to cost and efficiency. Ultimately, the resin and fiber selected for the composite are based on the structural and manufacturing requirements of the application. As such, the inherent chemical and rheological nature of the material must be understood. Through the work presented in this chapter, we established that for composites manufactured with an OOA process, it is critical that the material be used within its vendor-recommended out-time. When manufacturing thick composites of relatively small scale, we found that while a through-thickness temperature gradient exists during cure, material properties equalize following post-cure. For applications requiring material performance not met by the state-of-the-art materials, materials development efforts may enable PMCs to be used in new applications. This was demonstrated by incorporation of a thermoplastic veil at ply drop locations within a subscale fan blade component. Overall, contributions to the chemistry of materials and materials processing enabled application of composite materials for aerospace structures.

References

1. S.G. Miller, J.K. Sutter, D.A. Scheiman, M. Maryanski, M. Schlea, Study of Out-Time on the Processing and Properties of IM7/977-3 Composites. Proceedings of SAMPE Conference, Seattle (2010)
2. J.K. Sutter, S.W. Kenner, L. Pelham, S.G. Miller, D.L. Polis, C. Nailadi, T.H. Hou, D.J. Quade, B.A. Lerch, R.D. Lort, T.J. Zimmerman, J. Walker, J. Fikes Comparison of Autoclave and Out-of-Autoclave Composites. Proceedings of SAMPE Conference, Seattle (2010)
3. R.F. Handschuh, G.D. Roberts, R.R. Sinnamon, D.B. Stringer, B.D. Dykas, L.W. Kohlman, Hybrid gear preliminary results-application of composites to dynamic mechanical components. NASA/TM- 2012-217630 (2012)
4. S. Yi, H.H. Hilton, Effects of thermo-mechanical properties of composites on viscosity, temperature and degree of cure in thick thermosetting composite laminates during curing process. J. Compos. Mater. **32**(7), 600–622 (1998)
5. T.A. Bogetti, J.W. Gillespie, Process-induced stress and deformation in thick-section thermoset composite laminates. J. Compos. Mater. **26**(5), 626–660 (1992)
6. R. Joven, B. Minaie, Thermal properties of autoclave and out-of-autoclave carbon fiber-epoxy composites with different fiber weave configurations. J. Compos. Mater. **52**(29), 4075–4085 (2018)
7. R.J. D'Mello, M. Maiaru, A.M. Waas, Virtual manufacturing of composite aerostructures. Royal Aeronautical Soc **120**, 61–81 (2016)
8. Tencate Technical Datasheet, https://www.tencatecomposites.com/media/0b1de3c3-f3dd-4c01-b5c5-665919a00853/F4FQew/TenCate%20Advanced%20Composites/Documents/Product%20datasheets/Thermoset/UD%20tapes%20and%20prepregs/TC380_Epoxy_PDS.pdf
9. S.G. Miller, K.M. Handschuh, M.J. Sinnott, L.W. Kohlman, G.D. Roberts, R.E. Martin, C.R. Ruggeri, J.M. Pereira, Materials, manufacturing and test development of a composite fan blade leading edge subcomponent for improved impact resistance. NASA/TM-2015-218340 (2015)
10. T.K. Tsotsis, Interlayer toughening of composite materials. Polym. Compos. **30**(1), 70–86 (2009)

Author Biographies

Dr. Monica Allen is a senior research electronics engineer in the Munitions Directorate at the Air Force Research Laboratory (AFRL), Eglin Air Force Base, Florida. Her research focuses on resonant sensing platforms for electro-optical/infrared (EO/IR) technology components that can find applications in detection and sensing and micro-/nano-electronic-photonic hybrid circuits. Her primary areas of interest include nonlinear optical, photonic, and plasmonic physics concepts through the exploitation of light-matter interactions, while her other research interests include investigation of nonlinear optical phenomena and quantum photonic devices at infrared wavelengths. Her current projects range from modeling, simulation, and fabrication of micro- and nanostructures for gain and selectivity enhancement to characterization of fabricated devices to test their performance. She spent several years in industry working on highly sensitive sensing systems based on optical arrays and magnetostrictive materials. In the past, she has worked at the AFRL Sensors Directorate at Hanscom and Wright-Patterson Air Force Bases. She completed her PhD in Electrical Engineering from the University of Texas at Arlington and is affiliated with the Electrical Engineering Honor Society, Eta Kappa Nu; the Engineering Honor Society, Tau Beta Pi; and the Science and Engineering Research Honor Society, Sigma Xi and as a senior member of IEEE.

Monica is the youngest of three daughters who all pursued graduate degrees in science and engineering. Her parents raised their daughters to believe they could do anything. That's the kind of deep support that propelled her to the top of her field. Inspired early to use her imagination and Math skills to solve problems, she was captivated by any project that involved light, from beautiful diffraction patterns to reflections in a kaleidoscope. Today, the study of light is the centerpiece of her mission to devise groundbreaking optical technologies for the Air Force. As a principal investigator for AFRL, she plays a vital role in advancing state-of-the-art munitions

© Springer Nature Switzerland AG 2020
M. E. Kinsella (ed.), *Women in Aerospace Materials*, Women in Engineering and Science, https://doi.org/10.1007/978-3-030-40779-7

and weapon systems. As a Senior Research Electronics Engineer for the Munitions Directorate, she leads a research team of Air Force and academic collaborators that push the boundaries of technology to take on critical security and defense challenges.

Among other notable achievements, she has co-founded and led an international IEEE conference, called Research and Applications of Photonics in Defense (RAPiD), which drew more than 300 attendees from more than 10 countries. She also received the 2017 STEM Award in the senior investigator category at the Black Engineer of the Year Award (BEYA) Conference, a coveted award that reflects a lifetime achievement. She is a strong advocate for career development and has mentored several scientists and engineers. She is active in educational outreach, including teaching elementary students about physics, mentoring interns in the AFRL Scholars Program, and serving as an academic advisor for students earning graduate degrees. She truly values the importance of mentorship in her success, giving credit to her parents as early influencers, since they inspired imagination and a love of science and art. In school, her research and academic advisors were also inspiring and influential, further helping her to build self-confidence. Today, Monica's husband, Jeffery, is her biggest supporter. She considers herself to be part of the Air Force family and revels in the freedom to pursue extraordinary research opportunities while balancing the demands of a vibrant family life. She is realizing her dream to be a scientist, wife, and mother and is proud to work on our nation's most critical security and defense problems at AFRL.

In her spare time, she is an avid runner and competes in at least one half marathon a year. She is a voracious reader of both fiction and nonfiction and a movie enthusiast. She enjoys painting with watercolors and mixed media, including acrylics. Her favorite pastime is going to the beach or kayaking in the beautiful waters of the Gulf of Mexico, adjoining bays, bayous, and estuaries.

 **Angela Campo** is a research chemist at the Materials and Manufacturing Directorate, Air Force Research Laboratory (AFRL), Wright-Patterson Air Force Base, Dayton, Ohio. Her current research interest involves utilizing nuclear magnetic resonance (NMR) spectroscopy to optimize biomaterials production techniques. Working for AFRL has given her the opportunity to explore many research areas, adding to her technical knowledge and preparing her for future high-impact projects.

She has worked for AFRL since 2000 and began her career as a student researcher in the Fluids and Lubricants group. As she progressed through her career, she had the opportunity to work in several materials research areas. As a young chemist, she spent a significant portion of her career working with Lois Gschwender and Ed Snyder and was introduced to the world of aerospace hydraulic fluids, engine oils, greases, deicers, coolants, and cleaning solvents. She enjoyed learning about the complex relationships each of these materials has with aerospace systems, and those lessons have served her well throughout her career. During her time with the Fluids and Lubricants Team, Angela saw multiple products she researched be transitioned to the field. She helped champion the use of a corrosion-resistant grease (MIL-PRF-32014) in aerospace applications, conducted research and validation testing on the purification of hydraulic fluid, and performed custom synthesis of novel lubricants and additives for aerospace applications. Throughout her career, she gained experience in thermal management, novel thermoelectric materials, human thermoregulation, corrosion, coatings, and biomaterials research. She was the lead Air Force author for the Tri-Service effort to reinstate MIL-STD-1568 as an active military standard. MIL-STD-1568 enables the Department of Defense to establish corrosion requirements contractually, ensuring the best practices in corrosion prevention will be used during aerospace system design and sustainment activities. She is known for her ability to solve problems in the field ranging from unknown materials identification to rapid development of new materials or products to solve an urgent aerospace need.

Angela grew up in a small town in Ohio with her parents and two brothers. Her family had always been supportive of her scientific endeavors. Her parents encouraged her through many science fair projects, even letting her grow bacteria in her bedroom in a borrowed incubator. She was blessed to grow up in a home that believed women could do anything. Despite having many scientific interests, she realized that Chemistry was the best fit for her and was drawn to the idea of chemical synthesis and being able to produce compounds of interest. She obtained her BS in Chemistry from Wright State University (WSU) in 2001. After she joined the nanoscale thermal transport group, she decided to pursue a graduate degree. She graduated from WSU in 2010 with an MS in Chemistry. Upon joining the biomaterials research area, she decided once again to go back to school. Angela is currently a PhD candidate at WSU in the Biomedical Science PhD program. She is expecting to graduate in 2020. Working for AFRL has been a dream come true. She is able to pursue many research interests across multiple scientific disciplines while also helping the Air Force mission. Each day is different with new and exciting problems waiting to be solved.

She enjoys spending time with her husband, Darin; 14-year-old twins, Joseph and Eve; and her dogs, taking her kids to the wonderful metro parks in Dayton and introducing them to different crafts, traveling to interesting places, and renovating their home. Angela is also active in her church and loves to participate in volunteer opportunities with her children. She also participates in outreach to young scientists. For example, she has presented to a Women in Science and Engineering group at a local school, was a judge in the AFRL University Challenge on human cooling capabilities, mentored high school and college students, and judged local and district science fairs.

Dr. Joy E. Haley is a key leader in the development of optical materials for personnel, tactical, and space systems, specifically countermeasure products for the Air Force Research Laboratory's Materials and Manufacturing Directorate and the Department of Defense. She is an outstanding technical leader and senior chemist in the Photonic Materials Branch, leading a diverse team of scientists and engineers in organic nonlinear optical materials. In this role, she is responsible for setting the overall technical strategy, research quality, and resource execution/allocation in this area. For more than 15 years, her leadership has influenced not only the research agenda within AFRL but also those within the Navy, Army, and US international government partners. Recently in 2017, she became the research leader for the Nonlinear Electromagnetic Materials and Processes Research Team within the Photonics Materials Branch where she oversees a team of 10 government employees and 27 contractors including postdocs and students.

She grew up in Western Maryland where she developed a love for science. Early on, she was interested in paleontology and digging up dinosaur bones. Later, her interest evolved to paleoanthropology and early man. This sets the stage for pursuing a bachelor's degree in Biology from Frostburg State University. After enrolling in Chemistry as part of the course requirements, she quickly realized that obtaining a degree in Chemistry was more interesting and switched majors during her second year. There were certainly ups and downs going through the required courses. It was only after she took Analytical Chemistry that "she found her niche" as described by Professor Don Weser. During her time at Frostburg State, she was very involved in community outreach through the student-affiliated chapter of the American Chemical Society and served as president for 2 years. As her senior year started, it was time to figure out the next chapter in her life. She sent resumes out and applied to graduate school as well. In the spring, she decided to attend the University of Maryland Baltimore County (UMBC) for her master's degree in Chemistry. She finished her Bachelor of Science degree in May 1996.

With her interest in Analytical Chemistry, she started her first year at UMBC intending to pursue research in this area. However, things did not quite work out that way. She was given a desk in the laboratory of a new assistant professor, Dr. Lisa Kelly, who was doing work in the area of photochemistry/physical chemistry. After discussing potential research in developing novel chromophores for treatment of cancer through photodynamic therapy, she was hooked. As research and classes progressed, Dr. Lisa asked Joy if she would be interested in staying to pursue a PhD. She decided to go that route and received her PhD in Chemistry in January 2001. Dr. Lisa served as a constant mentor to her and provided the basis needed to be successful in her career. Having a woman role model was a key to her success.

In 2001, she moved to Dayton, Ohio, where she took a postdoctoral position working in the Air Force Research Laboratory on nonlinear optical chromophores. Since she was at first unfamiliar with nonlinear materials, she utilized her background in Physical Chemistry to fully understand these materials and become an expert in the field. She decided to stay as an on-site contractor and continue the work that she had begun. In 2010, she was hired by AFRL as a government civilian scientist. Soon, her job progressed from bench scientist research to directing the research of a team. Following her interest in outreach, she has been involved in the Dayton Chapter of the American Chemical Society and AFRL-sponsored outreach programs, hoping to excite the next generation of scientists.

Dr. Emily M. Heckman is a senior electronics research engineer at the Sensors Directorate, Air Force Research Laboratory, Wright-Patterson AFB, Ohio. She is the lead for the Microelectronics Integration Technology program and the research team lead and founder of the *Printronics* program, which focuses on the fabrication and testing of electronic and photonic devices using additive manufacturing techniques, such as inkjet and aerosol jet printing. She was instrumental in establishing printed electronics research at the Sensors Directorate and has built up a multimillion-dollar research lab for this work. The motivation of the *Printronics* program is to explore innovative direct-write fabrication technologies that will provide rapid prototyping, agile response, and conformal electronic devices to US warfighters. She is recognized as a leader in the area of printed electronics and has given numerous invited talks in the field. She has over 50 publications including several book chapters and is a senior member of both SPIE and IEEE professional societies. In 2017, she received the Affiliate Societies Council Outstanding Engineers and Scientists Award in Dayton, Ohio.

She earned her Bachelor of Science degree in Mathematics and Physics from the University of Dayton, her Master of Science degree in Physics from the University of Michigan, and her PhD in Electro-optical Engineering from the University of Dayton. Prior to her work in printed electronics, she was the Sensors Directorate team lead for the *Biotronics* program, a multi-directorate strategic technology thrust focused on integrating biomaterials into photonic devices. She established several biomaterial processing protocols and demonstrated improved performance using biomaterials in electro-optic devices as compared to the industry standard.

Growing up, she had little interest in pursuing a career in science or engineering as it took time away from her true passion: the theater. She had plans to be an actress and had the lead in almost every drama and musical performance throughout high school. However, despite her indifference to science, she maintained a strong interest and proficiency in Math, going so far as to take calculus courses at a nearby college while still in high school. This helped her tremendously when she decided, during her senior year of high school, to instead reach for a different type of star and pursue becoming an astronaut by way of studying astrophysics. After attending college summer internships in astrophysics at Harvard and the University of Chicago, she realized that astrophysics wasn't the type of hands-on challenge that she enjoyed (although it was fun to discover a galaxy while sifting through quasar data at the University of Chicago), and she instead pursued a career in electro-optics. She still has the opportunity to flex her acting skills, however, and uses them frequently in the pursuit of funding for her research.

Through STEM outreach talks, she uses her own background story of changing career fields from theater to physics, transferring schools both in undergraduate and graduate schools, and pursuing new career directions by jumping from optics to printed electronics to assure students that it's never too late to pursue their passions and that you can be successful no matter where life takes you. STEM education is very important to her. She has volunteered to talk with local schools and Girl Scout troops, lead tours of her laboratory, and help with Air Camp. Additionally, she mentors both undergraduate and graduate students through internships in her laboratory at the Sensors Directorate.

She is married to her high school sweetheart and is a busy mother of four children ranging in ages from 11 years to an infant. She enjoys running, reading, and lots and lots of coffee. She also enjoys traveling and loves the opportunity to expose her children to new places and experiences, especially if those places involve a beach! With the birth of her first child, she switched to a part-time working schedule (32 hours/week) and has maintained this schedule throughout her career. By taking advantage of alternative working opportunities like part-time hours and telecommuting, she has found a balance that still allows her to have a fulfilling and impactful career while raising a family.

Dr. Mary E. Kinsella recently retired from the Air Force Research Laboratory, where she served as a senior manufacturing research engineer in the Materials and Manufacturing Directorate at Wright-Patterson AFB, Ohio.

She lived in several places in the United States and Canada as a young girl. When she was 9 years old, her family settled in Southeast Michigan. Her parents ran a lumber business there and managed to send all their children to college. The middle child of seven, she was always a good student and did well in Math and Science. Engineering would have been a good choice for her undergraduate studies, but Mary did not know about engineering, and high school counselors did not encourage her to pursue it. Instead, she began studies in architecture at Miami University in Ohio. She soon found out that architecture was not for her and transferred to applied science to study manufacturing technology, which better suited her interests and talents. She earned her bachelor's degree and found a job in the microelectronics industry, where she worked in the clean room in integrated circuit processing. After gaining experience in supervision and materials control, she wanted a more technical position, so she began taking engineering classes. Then, when her husband-to-be was hired at the Air Force Research Laboratory (AFRL), she followed him there and found a job in manufacturing technology.

She began her career at AFRL working on electronics programs as a project engineer. She managed multimillion-dollar efforts in microelectronics manufacturing and commercial-military integration. More and more, she enjoyed the challenges in working on leading-edge technologies in the area of manufacturing and in her role as a government engineer. She was fortunate to be able to take advantage of higher education benefits and earned a master's degree in Materials Engineering from the University of Dayton and a PhD in Industrial and Systems Engineering at the Ohio State University. After receiving her PhD, she served as the chief of the Metals Processing Section and managed the Metals Affordability Initiative, implementing new materials and process technologies across the aerospace metals supply chain. She also served as the Additive Manufacturing Integrated Product team leader, responsible for bringing AFRL up to speed on 3D printing technologies and ensuring materials integrity and affordability for Air Force and DoD systems. As a founding member of the ASTM International Committee on additive manufacturing standards, her impact extended globally. Some of her other roles included the assistant chief scientist for the directorate, supporting strategic planning and execution of technical efforts for advanced acquisition and sustainment programs, and the AFRL university relations manager, responsible for outreach to and recruiting from prestigious science and engineering institutions across the United States.

She is a fellow and life member of the Society of Women Engineers, which has had a strong influence throughout her career. She is passionate about bringing more women into engineering and science and encouraging their advancement into leadership positions. She is now practicing as a professional coach to help women flourish in their careers as engineers and scientists.

She and her husband have navigated the challenges of a two-career family and raised two children, a daughter and a son, now working in professional fields. In addition to her coaching practice, she is a member of the advisory council for the Mechanical and Manufacturing Engineering Department at Miami University and serves on the board of directors for the Entrepreneurs Center in Dayton, Ohio. She enjoys classical music, yoga and wellness, needlework, and puzzles.

Dr. Sandi G. Miller is a chemical engineer at NASA Glenn Research Center, which is an amazing job that was not at all on her radar as a child. Growing up, she lived in North Haven, Connecticut, with her parents and one brother. During that time, her interests were wide ranging and primarily included art, photography, and animals. She did not know what she wanted in terms of a career but had considered veterinary school or advertising. Never had she imagined working for NASA on materials development for air and spacecraft. She moved to northeast Ohio after graduating from high school and attended Kent State University, on a pre-veterinary medicine track. She quickly learned that she had little interest in Biology; however, having enjoyed Chemistry courses at Kent, she decided to change her major. While in college, she developed a love of learning that had been absent during her high school years and strived to get as much from the college experience as possible. She was accepted into the Honors College and sought internships in her field. During her sophomore, junior, and senior years at Kent State, she worked at the Liquid Crystal Institute and completed an undergraduate thesis titled "Synthesis and Characterization of Cholesteric Polymer Liquid Crystals." The thesis was a requirement of the Honors College, but more importantly, the experience provided her an opportunity to integrate within a research group, understand how to perform research, and present her work to an audience.

After graduating from Kent State with a Bachelor of Science in Chemistry, she continued her education within the Macromolecular Science and Engineering Department at Case Western Reserve University in Cleveland, Ohio. She had intended to continue working in the field of polymeric liquid crystals; however, this did not pan out due to research funding constraints. Fortunately, she joined Professor Hatsuo Ishida's group, where her work was focused on nano-silicate clay dispersion in polymeric matrices.

Following graduation from Case Western in 1999, she was hired at NASA Glenn Research Center in Cleveland, Ohio. Her early work at NASA expanded upon the nanocomposite research she had completed in graduate school, although within NASA the research was focused on the development of polymer/nano-materials composites for aeronautics and space applications. More recently, her research priorities have shifted to include materials and processing of carbon fiber/polymer composites. As a subject matter expert, she has had the privilege to participate in multiple review panels within the composites community and mentor student interns throughout the year.

Working at NASA provided an opportunity to further her education, earning a PhD in Polymer Science from the University of Akron in 2008. During this time, she was married to Kyle Miller (2005) and had two children; Ashley, born 2006, and Claire, born 2008. Finding a balance between work and family life has not always been easy for her, but it has always been a team effort with support from her family and colleagues.

Outside of work, Sandi enjoys numerous activities that include traveling, CrossFit, and a wide range of sports, from golf to dirt bike riding and nearly everything in between.

Dr. Dhriti Nepal is a research materials engineer in the Materials and Manufacturing Directorate at the Air Force Research Laboratory (AFRL), Wright-Patterson Air Force Base, Dayton, Ohio. She directs research on controlled polymer composites processing, designing novel multifunctional composites, developing nanoscale characterization for materials performance evaluation, understanding chemo-mechanical effects and structure-property-performance evaluation, and guiding in-house multiscale modeling efforts. She has led and supported a number of interdisciplinary teams studying next-generation multifunctional nanocomposites, the optical properties of plasmonic nanoparticles for the next generation of sensors and optical communication devices, development of nanoparticles-based liquid crystals, biologically active nanocomposite thin films, and fibers. She is a recognized expert on directed assembly of nanoparticles and processing of polymer nanocomposites. She continues to be active in research as evidenced by 12 patents, 40 peer-reviewed scientific publications, 5 book chapters, 20 invited talks, and over 50 conference presentations throughout her career. She received the "Promising Professional Award" in 2018 from the Society of Asian Scientists and Engineers (SASE).

She influences others with her technical vision and serves as a role model to many, having mentored over 30 junior scientists, students, and visiting faculty. She also has numerous collaborations with universities and industry partners. Her research team has recently received awards in major international conferences such as the Materials Research Society (2018) and the American Chemical Society (2018). She seizes opportunities to inspire students into STEM careers, such as providing lab tours to Girl Scouts.

She was born in Kathmandu, Nepal, as a middle daughter of three girls in an incredibly supportive and loving family. She was passionate about science from her elementary school days and dreamed to pursue a PhD degree at some point in her life. As a teenager, she was inspired by congratulatory letters published every day in the local newspapers for the recent PhD graduates, who were considered as the elites of the society. Since Himalayan country Nepal did not have opportunities for higher education at that time, she went to South Korea to follow her dream with her husband. She was introduced to the new world of Materials Science and Engineering (MSE) in Korea by Professor Kurt E. Geckerler. She fell in love with the field and started its exploration to fulfill her dream, subsequently earning Master of Science (2002) and PhD (2006) degrees in MSE from Gwangju Institute of Science and Technology. She moved to the United States and studied polymer nanocomposites as a postdoctoral fellow at Auburn University (2006–2008), under Professor Virginia Davis, and Georgia Institute of Technology (2008–2009), under Professor Satish Kumar. With strong support from these advisors and mentors, she received another opportunity to work on the nanomaterials with world-renowned experts, including Dr. Richard Vaia at AFRL. The outstanding mentorship, research environment, and exceptional collaborative culture at AFRL provided a fertile platform for her research and career goals. She joined AFRL as a civil servant in 2015 and continued the exploration of nanocomposites for aerospace applications.

Outside of work, she is passionate about cooking, hiking, and social work.

Dr. Shanèe D. Pacley is a research materials engineer in the Materials and Manufacturing Directorate at the Air Force Research Laboratory (AFRL), Wright-Patterson Air Force Base, Dayton, Ohio. She is responsible for pulsed laser epitaxy growth of oxide and ultra-wide band gap materials for electronic devices. She specializes in high-quality epitaxial beta gallium oxide (β-Ga_2O_3) materials for power switching devices and ferroelectric materials for artificial intelligence. She also has years of experience working with superlattices for infrared detectors, synthesis and characterization of graphene and carbon nanomaterials for electronic devices, and two-dimensional materials, such as MoS_2, for low-power electronics. Currently, she has 17 peer-reviewed scientific publications and has been an invited speaker at conferences and in academic forums. She has also served SPIE, the international society for optics and photonics, on the organizing committee for their Photonics West conference sessions. As a result of her early-career accomplishments, she received the Black Engineer of the Year Award for Most Promising Engineer, the Women of Color in Technology Award, and the Federal Employee of the Year Award. She has participated in mentor programs, giving students the opportunity to gain hands-on experience in the research lab.

She was born in Dayton, Ohio, not too far from Wright-Patterson Air Force Base. As a child, she believed she would be a lawyer one day. However, after her first summer at a college prep program known as Wright STEPP (Wright State University Engineering Preparation Program), she knew she was going to be an engineer. She attended Wright STEPP for four summers and earned a full 4-year scholarship to Wright State University as a result. As an undergraduate, she participated in engineering organizations, such as the National Society of Black Engineers. Through the organization, she was able to participate in servicing the community, tutoring the youth, and helping at science fairs. Toward the end of her junior year, she attended a career fair and interviewed with the AFRL Materials and Manufacturing Directorate. She began her very first hands-on experience as a coop working with superlattice materials. Under the supervision and mentorship of Dr. Gail Brown, she learned a lot about type-II superlattice materials, such as InAs/GaSb, for photodetection. She continued working as a co-op and graduated with her Bachelor of Science degree in Materials Science and Engineering in 2003.

She then went on to graduate school at the University of Dayton. During this time, she continued as a co-op within AFRL, conducting research in superlattices and carbon nanotubes. The carbon nanotube research really piqued her interest and therefore became the focus for her thesis. She defended her thesis and graduated in 2006 from the University of Dayton with a Master of Science degree in Materials Engineering.

Following her graduation, she was hired as a government research materials engineer for AFRL. She married not too long after and was considering going back to pursue her PhD. Thanks to the encouragement and support of her husband, she enrolled and was accepted into the PhD program at the University of Dayton. Obtaining her doctoral degree, she believes, was the best decision she ever made. During those years, she learned about vacuum systems, epitaxy, catalysis, and more. In 2012, she received her PhD in Materials Engineering from the University of Dayton.

She continues to grow as a research materials engineer, presenting her work at conferences and workshops, writing papers, and collaborating with both academia and industry. She is adjunct professor at the University of Dayton, her alma mater, teaching both undergraduate and graduate courses.

**Nellie Pestian** is an undergraduate student in her third year pursing a Bachelor of Science in Materials Science and Engineering at Wright State University, Dayton, Ohio. She is currently working in the Composites Branch of the Materials and Manufacturing Directorate at the Air Force Research Laboratory as a student researcher under the guidance of Dr. Dhriti Nepal. Her research into polymer matrix composites thus far has included scanning electron microscopy (SEM) characterization of carbon fibers and cellulose nanocrystal thin films, as well as advanced nanoscale characterization with a focus on the interphase region of MoS_2-epoxy composites using SEM with energy dispersive X-ray spectroscopy, atomic force microscopy (AFM), and AFM with infrared spectroscopy.

She has always had a passion for Math and Science and attended multiple summer camps for women in engineering at Wright State University throughout her middle and high school years. When her high school Chemistry teacher, Elizabeth Hann, developed and executed a course for materials science, her interest in the subject matter grew. Mrs. Hann's passion for the topic inspired her to focus on materials science when the time came to pick a college major.

In her second year of college, she started looking for a way to get an internship at nearby Wright-Patterson Air Force Base. She knew that the experience from a research position would be invaluable for the development of her education and career, but she did not know where or how to start looking for a job. She learned about SOCHE, a contractor that employs many students at the base, from some of her peers, and filled out their online application. At first, she received no reply. Finally, a teaching assistant for one of her freshman engineering classes offered to recommend her to his supervisor on base, putting her on the candidate list. After a long process of interviewing and gaining approval to work in the secure government facility, she finally started to work in her current position.

She hopes to make connections through research collaborators and networking opportunities in her current position and upon graduation attends a graduate school that is highly ranked in Materials Science and Engineering where she can earn a master's degree and PhD. She enjoys the work that she does at the Laboratory and hopes to continue working in this field after finishing her education. She is especially interested in materials characterization. Many people suggest trying multiple internships to see what sort of work fits best, but she thinks that she got lucky in immediately finding work that interests and challenges her every day.

When she's not at work, she enjoys traveling and outdoor activities such as hiking and rock climbing, as well as playing team sports such as field hockey. Moreover, she likes to relax with a good fiction novel or a few episodes of *The Office*.

Rebecca Raig is an associate scientist under contract at the Materials and Manufacturing Directorate of the Air Force Research Laboratory (AFRL), Wright-Patterson Air Force Base, Dayton, Ohio. With a primary focus in microbiology, her current research revolves around optimizing cell-free protein synthesis (CFPS) as a means to produce intractable proteins of interest for the materials space. She became a full-time researcher at AFRL shortly after graduating from Miami University with a BS in Microbiology in 2018 and is motivated by the breadth of opportunity and experience presented to her at Wright-Patterson.

As an undergraduate, she worked as a student researcher in the Ferguson Structural Engineering Laboratory at Miami University, where she was closely mentored by Adam Creighbaum and Dr. D.J. Ferguson Jr. There, she gained a background in protein purification and pathway analyses from archaea and bacteria. This experience carried her student career away from Miami for two consecutive summers onto the Biological Materials and Processing Research Team at AFRL, where she learned numerous cultivation techniques for bacteria and fungi. Under the mentorship of Drs. Vanessa Varaljay and Caitlin Bojanowski, she was also responsible for assembling small-scale fuel tank environments in an effort to understand and counteract the costly biofouling issue facing fuel tanks across the Air Force. After her second summer as a student, she was once again able to embrace her roots in molecular biology, as she was invited to join the Synthetic Biology arm of the Biological Materials and Processing Research Team to begin her work in CFPS and related projects.

Outside of her research, she spends her free time teaching swing dance and mastering the perfect homemade cold brew. She was recently certified to teach fitness classes at a local studio and looks forward to taking part in the revitalization of downtown Dayton. Originally from Cleveland, Ohio, she is the youngest of four siblings that have simultaneously laughed at and supported her wanting to be a scientist by day turned professional swing dancer by night.

Dr. Amber Reed is a materials research engineer at the Air Force Research Laboratory's (AFRL) Materials and Manufacturing Directorate, Wright-Patterson Air Force Base, Dayton, Ohio. She has over a decade of experience in physical vapor deposition and characterization of thin films for electronic and optical applications.

She was born and grew up in Dayton, Ohio. She graduated from Eastern Michigan University with a Bachelor of Science in Engineering Physics with minors in Math and Chemistry in 2006. After completing her bachelor's degree, she started working toward a Master of Science degree in Physics at Wright State University.

During her first semester of graduate school, she worked in a terahertz spectroscopy laboratory.

She started at the AFRL Materials and Manufacturing Directorate as a student researcher in 2007 working on plasma treatments to increase the conductivity of indium tin oxide on temperature-sensitive substrates. She began her civil service career in 2009, first through the Student Cooperative Education Program and then as a pathway intern. During that time, she worked as part of a team on a variety of projects, including hybrid plasma sources for materials processing, energetic magnetron sputtering approaches, metallization of carbon nanotubes for thermal management, and high-power impulse magnetron sputtering (HiPIMS) of metals and oxides. Her dissertation was based on her work at the Materials and Manufacturing Directorate, in collaboration with the AFRL Sensors Directorate, on HiPIMS as a low-temperature, substrate-agnostic approach for depositing high-quality polycrystalline zinc oxide as the channel materials for thin film field effect transistors.

After completing her PhD in 2015, she was hired as a materials research engineer. Her current work focuses on physical vapor deposition of epitaxial transition metal nitrides for next-generation resilient tunable nonlinear plasmonics, low-loss piezoelectrics, and robust, oxidation-resistant electrodes.

In her free time, she is an avid reader of both fiction and nonfiction. She enjoys hiking and taking history tours and art classes.

Dr. Karen Taminger grew up in a Navy family, attending 11 schools before she graduated from high school. This experience helped shape her personality and her view of the world. For example, she *knew* that engineers worked in the hot boiler room in the lower deck of her father's ship. When she was in high school, she loved Chemistry and Math, but you can imagine she wasn't impressed when her guidance counselor suggested her to study engineering in college!

That all changed when she was selected to participate in the Virginia Governor's School, a 6-week summer experience at NASA Langley Research Center. There, her eyes were opened to the real world of engineering, from computers to wind tunnels, airplane wings to bubbling dewars of liquid nitrogen. She was also introduced to the concept of the cooperative education program, which would be her entrance into the NASA ranks as a 19-year-old college student from Virginia Tech. She started working at NASA Langley as a co-op student just 2 weeks before the Challenger accident, in January 1986, then came to work at NASA Langley full time in 1989 after graduating with her Bachelor of Science degree in Honors from Virginia Tech in 1989, majoring in Materials Engineering.

Throughout her career at NASA, she has worked in many different areas, all related to metallic materials for aerospace structures: fabrication, creep, simulated environmental exposures on mechanical properties of metal matrix composites, characterization of light alloys and metal matrix composites for elevated temperature applications, evaluation of structural performance of metallic sandwich and skin-stringer components, and X-ray diffraction. In parallel, NASA supported her pursuit of a Master of Science degree from Virginia Tech in 1999 in Materials Science and Engineering via distance learning. Since 2002, she has led NASA's development of the electron beam freeform fabrication (EBF³) technology, a large-scale metal additive manufacturing process for high-performance, low-cost fabrication of metallic structures for aircraft, launch vehicles, and spacecraft. One of the highlights of her career was the opportunity to conduct parabolic flight testing of EBF³ for compatibility with the space environment. As a result, she has spent 3 hours in zero gravity (in 15-second increments!).

She is now a senior materials research engineer at NASA Langley Research Center in the Advanced Materials and Processing Branch and currently serves as a technical lead for structures in transport aircraft and for several in-space and for-space manufacturing projects. She is the co-inventor on five issued patents and four other patent disclosures and coauthor on more than 35 papers and 100 presentations. She was awarded a NASA Exceptional Technology Achievement Medal in 2014 and a NASA Exceptional Achievement Medal in 2007. Her team was selected for the runner-up NASA Patent of the Year in 2016 and Langley's Whitcomb-Holloway Technology Transfer Award in 2008. All of these awards are related to their work in metal additive manufacturing.

She is married to an aerospace engineer and has three sons. Balancing career and home life is always a challenge, which keeps life lively and interesting!

Jill S. Tietjen, PE , entered the University of Virginia in the Fall of 1972, the third year that women were admitted as undergraduates (after suit was filed in the US district court), intending to be a Mathematics major. But midway through her first semester, she became interested in studying engineering and made all of the arrangements necessary to transfer. In 1976, she graduated with a BS in Applied Mathematics minoring in Electrical Engineering (Tau Beta Pi, Virginia Alpha) and went to work in the electric utility industry.

Galvanized by the fact that no one, not even her PhD engineer father, had encouraged her to pursue an engineering education and that only after her graduation did she discover that her degree was not ABET-accredited, she joined the Society of Women Engineers (SWE) and for more than 40 years has worked to encourage young women to pursue Science, Technology, Engineering, and Mathematics (STEM) careers. In 1982, she became licensed as a professional engineer in Colorado.

She started working jigsaw puzzles at age 2 and has always loved to solve problems. She derives tremendous satisfaction seeing the result of her work, e.g., the electricity product that is so reliable that most Americans just take its provision for granted. Flying at night and seeing the lights below, she knows that she had a hand in this infrastructure miracle. An expert witness, she works to plan new power plants.

Her efforts to nominate women for awards began in SWE and have progressed to her acknowledgment as one of the top nominators of women in the country. Her nominees have received the National Medal of Technology and the Kate Gleason Medal; have been inducted into the National Women's Hall of Fame and state halls including Colorado, Maryland, and Delaware; and have received university and professional society recognitions. She believes that it is imperative to nominate women for awards – for the role modeling and knowledge of women's accomplishments that it provides for the youth of our country.

She received her MBA from the University of North Carolina at Charlotte. She has been the recipient of many awards including the Distinguished Service Award from SWE (of which she has been named a Fellow and is a National Past President) and the Distinguished Alumna Award from the University of Virginia and the University of North Carolina at Charlotte and has been inducted into the Colorado Women's Hall of Fame. She sits on the boards of Georgia Transmission Corporation and Merrick & Company. Her publications include the bestselling and award-winning book *Her Story: A Timeline of the Women Who Changed America*, for which she received the Daughters of the American Revolution History Award Medal. Her most recent book, released in 2019, is *Hollywood: Her Story, An Illustrated History of Women and the Movies*.

Index

Printed in the United States
by Baker & Taylor Publisher Services